Glaciers

MICHAEL HAMBREY
JÜRG ALEAN

Glaciers

Published by the Press Syndicate of the University of Cambridge
The Pitt Building, Trumpington Street, Cambridge CB2 1RP
40 West 20th Street, New York, NY 10011-4211, USA
10 Stamford Road, Oakleigh, Melbourne 3166, Australia

First published 1992
Reprinted 1994

Printed in Hong Kong by Wing King Tong

A catalogue record for this book is available from the British Library

Library of Congress cataloguing in publication data applied for

ISBN 0 521 41915 8 hardback
ISBN 0 521 46787 X paperback

Frontispiece Mountain glacier descending steeply from Mount Foresta (3365 m) to join the fastflowing, heavily crevassed Hubbard Glacier in southern Alaska.

Contents

'Out of whose womb came the ice? And the hoary frost of heaven, who hath gendered it? The waters are hid as with a stone, and the face of the deep is frozen.'

Job, ch. 38, vv. 29–30

Preface

Glaciers are among the most beautiful and fascinating elements of nature. Slowly they creep and slide from mountain regions to the lowlands, and over millions of years they have been shaping the landscape by scouring away the rocks, and transporting and depositing debris far from its source. In so doing, they have created some of the finest landscapes on earth. Moreover, they provide the meltwater that drives turbines and irrigates deserts, or furnish material for soils that is particularly suitable for agriculture. But glaciers can also destroy human property and take people's lives.

As glaciologists we have attempted to understand some of the infinite varieties of glacial phenomena. We have lived for months at a time on glaciers and have seen them in all their moods. They have often presented a benign appearance, as on a calm, sunny day, when travelling over them has been safer than walking the streets of a busy city. At other times, as when blinding blizzards have obliterated our paths and the snow has treacherously hidden the crevasses, glaciers have unnerved us and made us wish for the security of home. But time and again we have been drawn back to glaciers, eager to absorb their natural beauty as well as gain a better appreciation of how they behave.

In this book we wish to take the reader, at least in imagination, on excursions to glaciers, both in far distant and remote lands and in regions close to human occupation. First we describe several characteristics shared by most glaciers. We consider the role glaciers have played in a global context in the past and are playing at the present time (chapter 1). Glaciers come in all shapes and sizes, from a small patch of ice a few hundred metres across to the huge ice sheet that almost totally buries the continent of Antarctica, and these are described in chapter 3. Glaciers are not static: they are born and subsequently respond to increases and decreases in their supply of snow (chapter 4), and they flow, the subject of chapter 5, which also embraces the various structures that result from flow, as well as some of the spectacular events associated with unusual flow behaviour. Neither are they bare sheets of ice, but can transport large amounts of debris (chapter 6). The intimate association of ice and glacial meltwater is outlined in chapter 7, and that between glaciers and the sea, and the generation of icebergs in chapter 8. The violent and often disastrous interaction between glaciers and

Preface

The world's largest ice sheet, on Antarctica, is fringed in places by ice shelves—slabs of glacier ice floating on the sea. Here the edge of the Ekstroem Ice Shelf has a rough, knobbly surface because of the freezing of spray when waves lashed the seaward-facing cliff during a storm in March 1991.

volcanoes forms the basis of chapter 9. If this mix can create sudden changes to the landscape, even more important ones happen on a far slower time scale, as erosion and deposition are revealed once ice has retreated, giving glaciated landscapes their distinctive character. In a chapter on wildlife in glacial environments we consider how some animals and plants have adapted to the severe climatic conditions. The book closes (chapter 12) with a look at how glaciers and the human race interact with each other. We describe a number of localized events where glaciers have been responsible for major catastrophes, but also discuss the benefits that glaciers have brought to civilization. We also consider the climatic record of glaciers, and ice sheets, and the global implications if major changes in the ice cover were to occur.

The photographs have been selected in order to demonstrate some of these many facets of glaciers behaviour.

In summary, our aim is to describe and explain glaciers in all their variety, as well as the landscapes they are creating and are still creating, with illustrations taken mostly by ourselves in six continents. If we succeed in conveying to our readers the beauty and importance of glaciers, we will have fulfilled our purpose.

Michael J. Hambrey
Cambridge, England

Jürg C. Alean
Bülach, Switzerland

Acknowledgements

Michael Hambrey wishes to thank: his parents, who were the first to foster his interest in the glaciated landscapes of the highland areas of Britain; Wilfred Theakstone of Manchester University, who first inspired his interest in glaciers and guided him through his PhD studies on a Norwegian glacier between 1970 and 1973; Geoffrey Milnes, then of the Swiss Federal Institute of Technology in Zürich, for the opportunity to work with him on Alpine glaciers from 1974 to 1977; the late Fritz Müller of the same institution for opportunities to work on glaciers on Axel Heiberg Island in the Canadian Arctic in 1975; Brian Harland, of Cambridge University, for participation in geological expeditions to Svalbard between 1977 and 1983; Martin Sharp, also of Cambridge University for the chance to work on an Alaskan glacier in 1986; Peter Barrett of the Victoria University of Wellington for the invitation to join a research project in Antarctica in 1986; Niels Henriksen for facilitating fieldwork in East Greenland with the Geological Survey of Greenland in 1988; Dieter Fütterer, Gerhard Kuhn and Werner Ehrmann of the Alfred Wegener Institute for Polar and Marine Research in Bremerhaven (Germany), for arranging funds to pursue collaborative Antarctic research; and Victor Smetacek, Chief Scientist on a cruise of FS *Polarstern* to Antarctica, for facilitating photography of additional Antarctic scenes in 1991. In addition, numerous other geologists are thanked for leading field excursions in which the author has participated to various other glacier-influenced regions of the world and which figure in this book. Many of these photographs could not have been taken but for the skills of the pilots of chartered helicopters and fixed-wing aircraft, and the boat crews of both large and small research vessels.

In each of the above areas, as well as Greenland, the author's work (whether or not specifically glaciological) has been financed by the UK Natural Environment Research Council, the Swiss Centenarfond, the Royal Society of London, the Cambridge Arctic Shelf Programme, the New Zealand Antarctic Research Programme, the Alfred Wegener Institute, the Leverhulme Trust and the Transantarctic Association. The work has been assisted immeasurably by the successive hospitality of the Department of Geography, University of Manchester; the Swiss Federal Institute of Technology in Zürich; the Department of Earth Sciences and the Scott Polar Research Institute of the University of Cambridge; and the Alfred Wegener Institute.

Jürg Alean offers his thanks to his parents who first stimulated his interest in the natural environment, the late Eugen Steck who taught him how to observe accurately and patiently, and Peter Weber who showed him what good mountain photographs should look like. Paul Felber was partly responsible for Jürg Alean having studied earth sciences. Wilfried Haeberli provided guidance as his PhD supervisor and friend at the Swiss Federal Institute of Technology. Many other colleagues and friends deserve thanks for their academic and field support. Most important of all was his wife's understanding, patience and support during their joint global wanderings.

For permission to reproduce certain figures we would like to thank David Drewry for the drawing on page 25, from *Glaciological and Geophysical Folio* (published by the Scott Polar Research Institute); and the US Geological Survey, Flagstaff, Arizona, for the satellite image on page 32. Other figures have been modified from: Bruce Molina, *Glaciers of Alaska* (Alaska Geographic Society): Louis Renaud, *Journal of Glaciology* 1987; and C. Lorius and others, *Nature* 1985.

Finally we would like to thank Wilfred Theakstone (University of Manchester), Clifford Embleton (University of London) and two anonymous reviewers for making helpful suggestions on the manuscript. We thank also Peter Richards of the Cambridge University Press and Stephen Adamson, our editor.

1

Glaciers, Past and Present

Glacier ice covers as much as 10 per cent of the earth's land surface at the present time. In geological terms we are living in a glacial era that began in Antarctica perhaps 40 million years ago. However, the latter stages of this era have included many alternations between periods of full glaciation, when much of the northern hemisphere was covered by ice, and interglacial periods with much less ice, as at the present day. The most recent full-scale glaciation, when ice covered up to 30 per cent of the land, ended as recently as 10,000 years ago, and a return to such conditions may be expected in a few thousand years' time, unless the disruption to the earth's climate system by man is so severe that the enhanced greenhouse effect caused by environmental pollution wins.

Huge ice sheets have advanced across Europe and North America during a succession of glacial periods in the last two million years. A major ice sheet developed several times over the Scandinavian highlands and spread west across the North Sea to join a smaller British ice sheet. At its most extensive, ice covered the whole of Great Britain as far south as a line through Bristol and London. From Scandinavia the ice moved into northern Germany and Poland. At the same time, large glaciers fed by the mountain snows of the Alps spread out over the surrounding lowland. In North America ice from the Arctic and the Western Cordillera covered almost the whole of Canada and extended south over the prairies of the Mid-West, and the areas now occupied by Chicago and New York City vanished under the ice. In many other parts of the globe similar build-ups and retreats of ice took place, for example in the Andes, central Asia and New Zealand, and even Australia. In contrast, the Antarctic and Greenland ice sheets appear to have remained largely intact, although there is some debate as to whether the latter disappeared during the last interglacial period, about 130,000 years ago. The relatively small fluctuations of these ice sheets is due to iceberg production at the edge of the bordering continental shelves that prevented their expansion.

In the context of Earth's entire 4600 million year history, glacial periods have been relatively rare. Nevertheless, continental-scale glaciers have occasionally developed at different times on all continents. Even areas like the Sahara and the middle of Australia that are hot deserts today, and the tropical areas of Brazil, bear signs that glaciers

Many upland areas far removed from present-day glaciers show strong signs of glacial erosion. Glacial troughs with their steep, straightened sides and bedrock basins filled by lakes are typical, as in this view of the Buttermere Valley in the English Lake District. The two lakes, Buttermere in the foreground and Crummock Water, were once a single lake, but deposition of gravel by a stream entering from the right has divided the original single lake in two.

Central Park, New York, USA. During the last glaciation the North American ice sheet flowed as far south as New York City. Its effects are visible here in the scratched and rounded rock outcrops.

once were present, albeit hundreds of millions of years ago. At such times both the world climate and the relative positions of the moving continents were very different from those of today.

Excluding water in the ground, glacier ice represents 80 per cent of the world's fresh water, of which 99 per cent is locked in the ice sheets of Antarctica and Greenland, far removed from most human activity. However, in some countries, such as those of central Europe and Scandinavia, glaciers have affected the lives of people for centuries, not always to their benefit.

Glaciers are important for recreation, especially for climbers and skiers, and contribute greatly to the beauty of a mountain landscape.

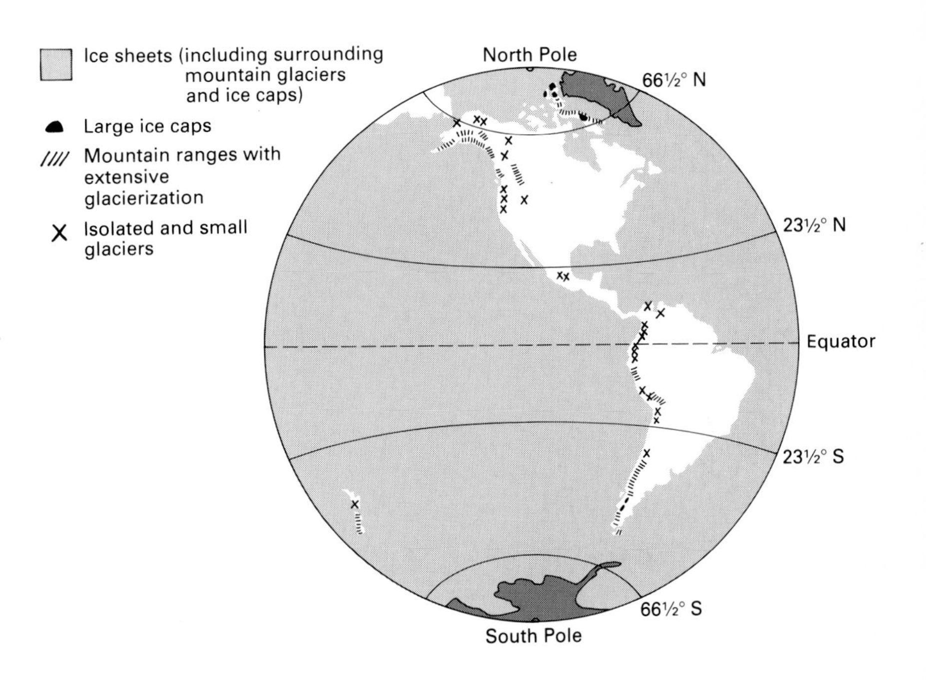

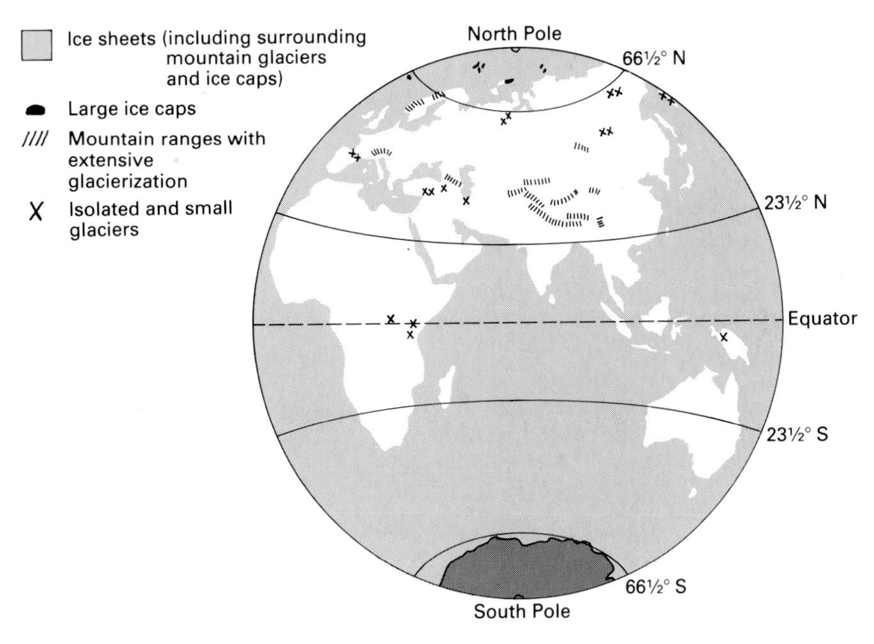

Present-day distribution of glaciers and ice sheets.

Below Deposition direct from glaciers creates a mixture of boulders, sand and mud called 'till'. If this is then buried and turned into a rock, a 'tillite' is produced. Here one around 650 million years old on C. H. Ostenfeld Nunatak in East Greenland became exposed when a contemporary glacier retreated. The scratch marks, or striations, are typical of glacially eroded surfaces, but were formed in the recent past.

Right The world's highest mountain chain, the Karakorum and Himalayas, carries a substantial volume of ice which flows to lower ground as valley glaciers. The ice erodes the high peaks, leaving steep ridges and pyramids. One such pyramid is K2 (Qogir Feng; 8611 m) in the Karakorum, the world's second highest peak (right of centre).

This striated stone in the western Sahara, Mauritania, was abraded at the base of an ice sheet that covered much of the surface of the earth about 650 million years ago.

They are also an important water resource, capable of supplying water throughout a long, hot, dry summer, and considerable amounts of glacial meltwater have been harnessed for the generation of hydro-electricity. In the future, Antarctic icebergs may even provide water for the parched regions of Africa, the Middle East or Australia. Yet glaciers can kill. An unwary or careless walker may fall through a snow bridge on a glacier into a hidden crevasse, or be crushed by an ice avalanche. Far from the glaciers themselves, large-scale loss of life has been caused by huge ice avalanches and floods of water that have burst unexpectedly from beneath a glacier, and valuable pastures and roads have been lost under advancing glacier tongues.

Glacial erosion and deposition have an equally important effect on human activities. The slopes of glacially eroded valleys can be so steep as to be prone to rockfall. Glacial deposits over lowland areas have not only provided some of the richest farmland, but have given us abundant reserves of sand and gravel for use in the construction industry, as well as local concentrations of valuable metals.

The large ice sheets are sensitive to climatic change, but in ways that we do not fully understand. Nevertheless, we need to take seriously the threat that global warming may have on sea levels, even though no clear relationship has been established between rising temperatures and ice sheet melting.

Some glaciers are tiny, merely a few hundred metres long, whereas the largest are represented by the great ice sheets of Antarctica and Greenland, and all shapes and sizes exist in between. Naturally enough, the glaciers that have received most attention from glaciologists are those in the mountain regions near the population centres of the USA and Europe. The earliest descriptions of glaciers date from the eleventh century in Icelandic literature, but the fact that they move was apparently not recorded until about 500 years later. The first serious scientific studies were in the late eighteenth century, and since then the subject has been tackled by increasing numbers of geologists, geographers, physicists, chemists, mathematicians and meteorologists, all of whom could also be considered glaciologists. Nowadays, glaciers are studied on remote Arctic islands, on the Antarctic and Greenland ice sheets, and of course on more accessible sites. As a result of all this activity our knowledge of them has improved considerably in the last forty years.

If anything the pace of this research is quickening; because of their importance as reservoirs of water and their sensitivity to climatic change, the world's glaciers are being systematically documented, and several countries have produced their own glacier inventories for the compilation in the near future of a World Glacier Inventory. This will provide for the first time an indication of the total volume of freshwater on the surface of the earth.

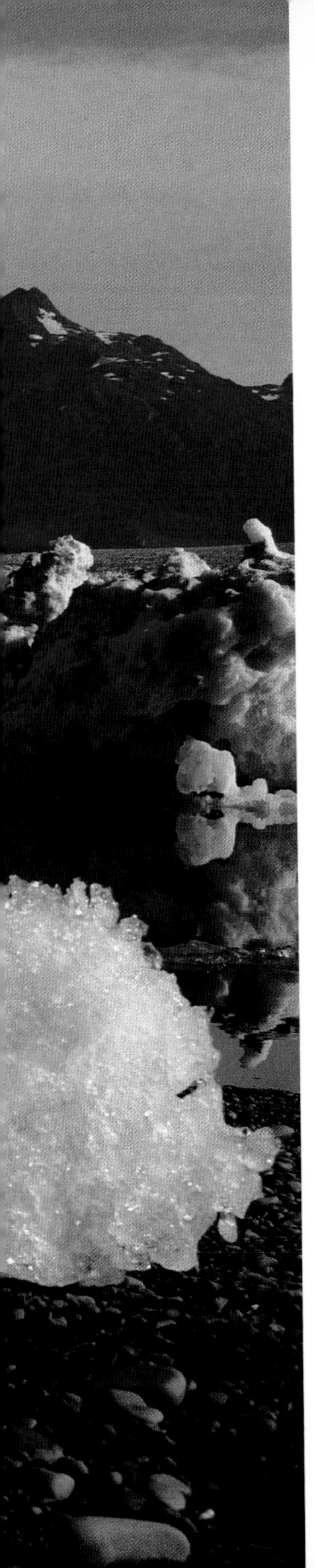

2

The Anatomy and Health of a Glacier

Glaciers are frequently thought of as frozen rivers, but this is not so. Most glacier ice has never been in the form of running water but is principally derived from snow. For a glacier to form, first the winter precipitation has to be great enough for some of the snow to last throughout the following summer. Then this is repeated for several successive years. Finally, under the pressure of its own weight the snow turns to ice. If this ice is thick enough, it can flow under the influence of gravity. This transformation of snow to ice is often a long and complex process, both its nature and the time involved depending on temperature and the depth of further, overlying snow. The changes are fastest in temperate regions, like the Alps and the Western Cordillera of North America, and slowest in the polar regions, because of the higher temperatures in the former.

The transformation of snow to ice

Although all snow crystals have a hexagonal structure, with characteristic six-sided symmetry, snow falls in myriad forms. Snowflakes may come as delicate, feathery crystals a centimetre or so across, or as relatively hard grains that have the feel of sand. They have their most intricate and varied forms when they fall close to freezing point, and can form a very light snow layer with a density of only a twentieth that of water. Snow like this on the ground is fluffy – the powdered snow much beloved by skiers.

A glacier can form where the annual accumulation of snow exceeds the amount melting or evaporating each year. In high mountains accumulation is related to temperature, which in turn is dependent on altitude. The amount of snow is also significant: wind leads to greater accumulation on the lee (downwind) side of a mountain than on the windward side. Also, understandably, the shady sides retain more snow.

Even meltwater has a bearing on the build-up of a glacier. In temperate areas, where meltwater is abundant, snow is converted to ice in a number of stages. First, the fragile crystals break on settling, as they are compressed by the weight of additional snow on top, or they break down if they become wet. Gradually, snowflakes change to grains which become rounded and granular, like coarse sugar. As the snow becomes compressed it becomes harder and denser. At first the air spaces

Previous page The Columbia Glacier in southern Alaska began a catastrophic retreat in the mid-1980s. At the time this photograph was taken in 1989 the glacier had retreated several kilometres, producing numerous icebergs.

The transformation of snow to ice involves the burial of successive annual layers of snow. In temperate and tropical mountain regions each annual layer may be several metres thick. The 5600-metre-high Wallunaraju in the Cordillera Blanca of Peru, towering behind these climbers, shows typical annual layering.

between grains are connected, but now the snow is said to have turned into 'firn' (a term derived from the German, meaning 'old snow'); this is an intermediate stage in its transformation to ice. The firn stage is generally reached after one complete winter–summer cycle, when the snow's density approaches a half that of water.

As these changes proceed, the relatively round grains of firn begin to recrystallize and larger crystals of ice begin to form at the expense of their smaller neighbours. Air is now only present as bubbles trapped inside the growing crystals. These changes are aided by the flow of the glacier, as ice crystals deform in a similar manner to plastic substances, and in a flowing glacier the form of the crystals is in constant change. If the ice is deforming rapidly the crystals may not grow very large because they never remain stable for long. However, by the time ice reaches the glacier terminus, where flow rates are slower, crystals may grow several centimetres long. In stagnant ice they may grow even larger, perhaps as much as 25 centimetres in length, acquiring a very complex shape. By this time ice has a density of around 90 per cent that of water.

In glaciers where there is no surface melting at all, as at high altitudes in Greenland or Antarctica, the change from snow to ice is very slow, taking hundreds of years rather than a few. The transformation here is determined by three factors: the movement of crystals relative to each other, the effect of increasing compaction under the accumulating snow, and internal deformation.

Observers are often struck by the blueness of some glacier ice, which is especially noticeable on cloudy days. This is because water molecules absorb all colours of the spectrum except blue.

Profit and loss

The changes of snow to ice and movement downslope are often seen in terms of a balance between profit (accumulation) and loss (ablation), in a similar way to that of a bank account. This characteristic is known as the mass balance or mass budget.

We can observe the changes from profit to loss by walking down a glacier in late summer. At the top, snow accumulation exceeds melting. This part of the glacier is called the accumulation area. Its upper part may consist of a zone of dry snow where there is no melting, but only polar or very high altitude glaciers have such a zone. Passing downwards we enter a zone where there is some melting. Water percolates through

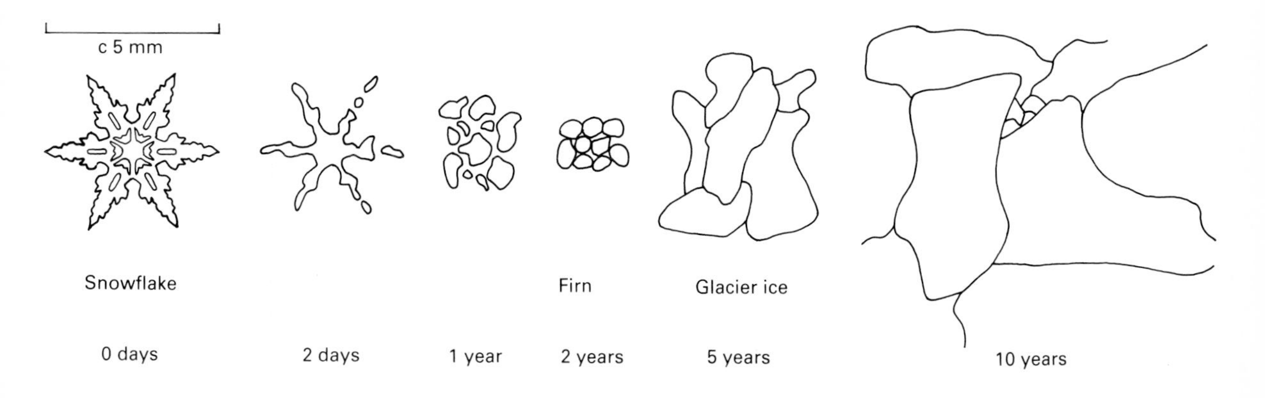

Transformation of snow to glacier ice crystals.

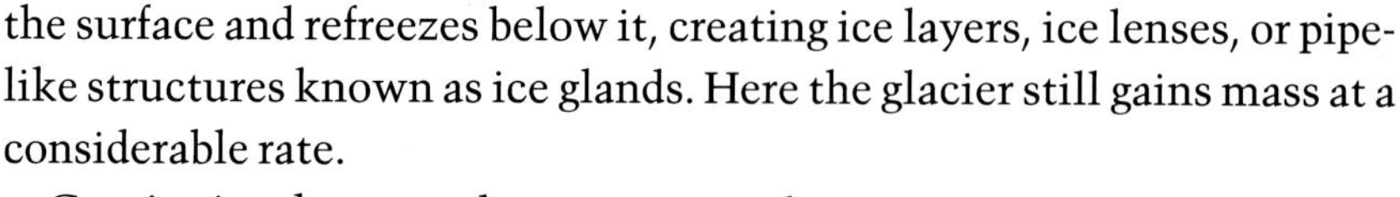

New Zealand's biggest glacier, the debris-covered Tasman, is rather inactive and is wasting downwards, rather than retreating.

the surface and refreezes below it, creating ice layers, ice lenses, or pipe-like structures known as ice glands. Here the glacier still gains mass at a considerable rate.

Continuing downwards we pass into the wet snow zone, where most, if not all, the gains have been lost. All the snow here is raised to melting temperature and becomes very wet. On polar glaciers large areas of slushy snow are formed in this area, much of which refreezes in winter to form superimposed ice, and some of which adds to the glacier's mass. The boundary between the superimposed ice and the wet snow zones is known as the firn line. On glaciers in more temperate regions, however, this superimposed zone is generally rather narrow. In contrast, in many parts of Antarctica this, and lower, zones do not exist as the dry snow zone extends to sea level where the ice 'calves', meaning that it produces icebergs that float away from the glacier margin.

The lower limit of superimposed ice represents the most important boundary: the equilibrium line, which in temperate regions coincides approximately with the firn line. Here profit equals loss. Below the equilibrium line, in what is known as the ablation area, expenditure exceeds income, and we find that all the previous winter's snowfall has melted, as well as an increasing amount of ice. In the lower part of a typical Alpine glacier the melting of the ice may exceed 10 metres a year, producing increasing mass deficit as one moves downwards. However, this does not necessarily mean that the glacier retreats, since the lost ice is normally replenished by ice flowing from above.

End-of-summer view of the upper part of the Norwegian glacier Charles Rabots Bre showing the boundary or 'firn line' between the preceding winter's snow and the older, dirty firn and ice.

In early summer in the ablation zone large areas of the melting snow pack become saturated with water, especially if the glacier's surface is flat. Such 'snow swamps' are more common on polar glaciers because cold ice (ice below melting point) stops meltwater draining away.

As summer progresses, large ice crystals melt along their boundaries in sunny or dry weather, so that the ice acquires a nobbly surface texture which makes it easy to walk on, even without crampons. In contrast, in rainy weather ice melts evenly and becomes much more slippery. Sunny weather is also prone to enhance the surface undulations, especially if there are uneven concentrations of debris, and can produce what are known as ice ships. These are often a metre high, but examples several metres in height are found where solar radiation is very strong.

Right Crystal from an iceberg that originated from the rapidly retreating Columbia Glacier, Alaska. Individual crystals may grow to a considerable size in relatively stagnant ice.

Below As snow turns to ice, irregular interlocking crystals develop and air bubbles are trapped. A thin slice of ice taken from the Norwegian glacier Charles Rabots Bre is viewed between a pair of polaroid plates; individual crystals are denoted by different colours and trapped air bubbles are also visible.

Long profile through a glacier depicting the accumulation and ablation areas, with the associated flow paths of material as it becomes buried and later emerges at the surface.

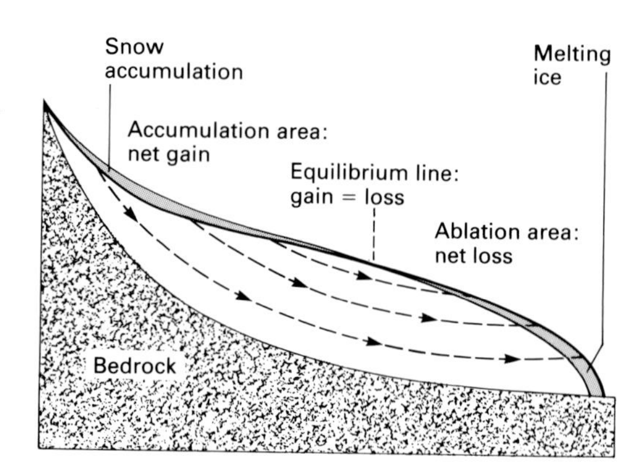

Mass balance changes

The overall or net balance between profit and loss is reflected in the response of the lower reaches of the glacier, that is its 'snout'. A 'healthy' glacier is one in which as much ice is formed in the accumulation area as is lost in the ablation area, in which case the snout remains in the same position. If gains exceed losses, the glacier will advance. An 'unhealthy' glacier loses more than it gains and its ice margin will recede (even if the ice continues to flow forwards), or the glacier will stagnate and waste downwards.

The Southern Alps of New Zealand provide a remarkable contrast in the response of glaciers to changes in mass balance. On the maritime west flank of the Mount Cook range precipitation over the Fox and Franz Josef glaciers reaches the phenomenal value of 10 to 15 metres a year of water equivalent, much of it falling as snow at high altitude. The glaciers descend into rain forest where it hardly ever snows, at velocities reaching an annual average of 700 metres a year, although in periods of heavy rain when the beds are lubricated they may attain 2500 metres a year. The attendant mass balance changes are reflected in a very rapid response of the glacier snouts, with advance and retreat rates of typically 0.5 to 1 kilometre a year.

A few kilometres across the Mount Cook divide the glaciers are much less dynamic. Here the precipitation drops off rapidly, so that near the snout of the Tasman Glacier it is only 0.4 metres a year. Although this glacier is the biggest in New Zealand, it has a maximum velocity of only

250 metres a year, and the snout does not leap forwards or backwards like the Fox or Franz Josef glaciers, but thickens or thins along its entire ablation area, often called the 'tongue' on account of its shape.

In the European Alps, and many other areas, it is possible to observe both advancing and retreating glaciers in the same valley under the same climate. How can this be reconciled with mass balance? Small glaciers tend to respond within a few years to changes in mass balance, whereas the larger Alpine valley glaciers like the Grosser Aletschgletscher may take half a century or more. For large glaciers, then, short-term variations in mass balance will not be reflected in changes in the snout position; they will only respond to longer-term changes in climate. The main exceptions to this rule occur in areas where the snowfall is exceptionally heavy, as on Fox and Franz Josef.

Generally, climatic changes over a period of several decades will counteract short-term variations. This is why slow climatic warming since the turn of the twentieth century has led to almost all the glaciers in North America and Europe retreating considerably, a pattern partially reversed in the late 1970s and early 1980s.

In contrast to valley glaciers, such as those referred to above, polar ice sheets and ice caps respond much more slowly to climatic changes. Indeed, the warm temperature wave that accompanied the end of the last ice age around 10,000 years ago has still not fully penetrated the main ice sheets, and it is thus unlikely that they are ever in equilibrium. Mass balance changes in them also only respond slowly. On many parts of the East Antarctic Ice Sheet precipitation rates are only a few centimetres a year, and ablation is predominantly by calving into the sea, rather than by melting. As the several ice drainage basins which make up the East Antarctic Ice Sheet cover up to a million square kilometres, changes in mass balance need hundreds of years to be transmitted to movements in the position of the ice margin. Furthermore, changes of the ice margin may be controlled more by oceanic factors, such as the temperature or level of the sea, than by changes in mass balance.

Longer-term climatic changes, over thousands of years or more, are reflected in huge changes in ice volume, and the comings and goings of ice ages. Here we are looking at astronomically induced variations in solar radiation input at the earth's surface.

Overleaf Outlet glaciers from two ice caps in southern Axel Heiberg Island, Canadian Arctic, descend into a valley where they spread out to form piedmont glaciers. The subdued surface features are accentuated by the low angle of the August midnight sun.

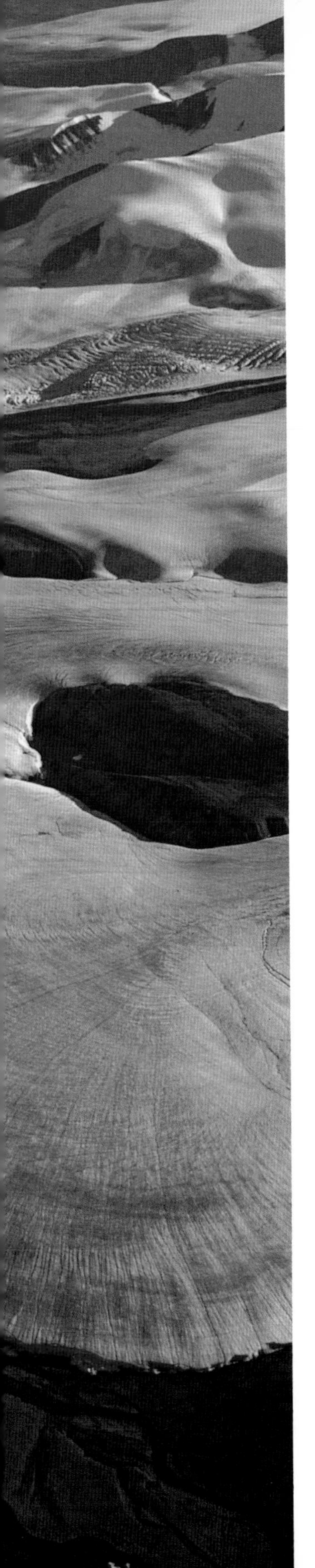

3
The Glacier Family

Glaciers are usually classified according to their shape and their relationship with the surrounding and underlying topography, but some are described on the basis of the temperature distribution within the ice. However, it should be borne in mind that these distinctions are not strict and that transitions exist between all these types of glacier.

Types classified by setting

Ice sheets and ice caps

The largest of the topographically defined glaciers are the ice sheets of Antarctica and Greenland. Covering a continent a third bigger than Europe or Canada, or twice as big as Australia, the Antarctic Ice Sheet contains 91 per cent of the world's freshwater ice and 85 per cent of its freshwater. It attains a thickness in excess of 4000 meters in places, inundating entire mountain ranges. In much of West Antarctica, ice rests on bedrock that is many hundreds of metres below present sea level. Consequently, if the West Antarctic ice were to melt it would be replaced by a sea with numerous island archipelagos. In contrast, the ice sheet over East Antarctica sits on relatively high ground, but drains via a number of buried valleys, some of which are below sea level. Apart from the Transantarctic Mountains and the mountainous backbone of the Antarctic Peninsula, which are high enough to project above the level of the ice sheet, rock outcrops, known as *nunataks*, a word of Inuit origin, are few and far between. The much smaller Greenland Ice Sheet, which contains 8 per cent of the world's freshwater ice, nevertheless covers an area the size of Mexico or ten times that of the British Isles. The Inlandis (Inland Ice), as it is known, fills a huge basin that is rimmed by ranges of mountains to a depth of over 3000 metres. Ice overflows and breaches this rim in many places, discharging into the sea and producing icebergs.

Ice caps are smaller versions of ice sheets, generally defined as covering an area of less than 50,000 square kilometres – many only a few square kilometres. Like ice sheets, they inundate the underlying topography, and their smooth surface belies the irregular nature of the bed. Ice caps are common in polar and sub-polar regions, but are rare in regions of alpine topography. They tend to develop on high plateaux from which discharge is inhibited by the low gradients, except where ice

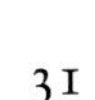

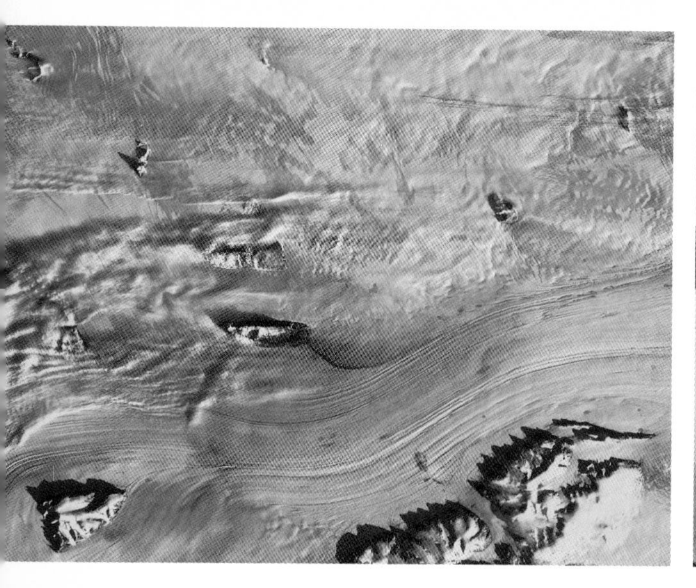

Far left Landsat satellite image of part of the East Antarctic Ice Sheet (the Lambert Glacier), showing bare ice areas where ablation exceeds accumulation.

Left Two small ice caps on southern Axel Heiberg Island in the Canadian Arctic show the typical radial flow pattern. Ice accumulates on a high area in the middle and spills down on all sides.

Left below The Greenland Ice Sheet fills a large basin bounded by a ring of coastal mountains. As the ice spills out of the basin and begins its descent towards the coast, it has to flow around the higher peaks, which project through the ice as 'nunataks'. This view was taken from Tillit Nunatak (2060 m) in central East Greenland.

Below Cross-section through the East and West Antarctic Ice Sheets, illustrating the irregular nature of the bedrock and the ice thickness, as well as floating ice shelves.

spills over the plateau edge. Among the largest are Austfonna and Vestfonna on Nordaustlandet in Svalbard, covering an area the size of Wales or Connecticut, and Vatnajökull in Iceland.

Both ice sheets and low-level ice caps discharge in a variety of ways; for example, through valleys on land, by breaking up as the ice flows off the cliff bounding the plateau, or directly into the sea. Those that flow into the sea have a unique feature – ice streams. These are zones of much faster flow which have well-defined boundaries with slow-moving ice; they behave like separate glaciers and commonly have a very crevassed surface, whereas the bordering ice is far less disturbed. Their margins are zones of very intense shearing, reflected in the development of deep, closely spaced but short crevasses. Some ice streams in Antarctica extend into the sea as unconstrained floating glacier tongues which periodically break off as huge icebergs.

Ice shelves

There is a net accumulation of snow and ice close to sea level in the Antarctic, which encourages the formation of ice shelves – slabs of glacier ice which float on the sea, and which typically range in thickness from over 1,000 metres in their inner parts to 500 metres or less where they calve. Those glaciers that discharge into the sea from higher ground become detached from the bed and float, spreading out to cover large bays, as happens in the Ross and Weddell seas. Some ice shelves cover vast areas; the Ross Ice Shelf, for example, measures about 850 by 800 kilometres and covers over half a million square kilometres, which is

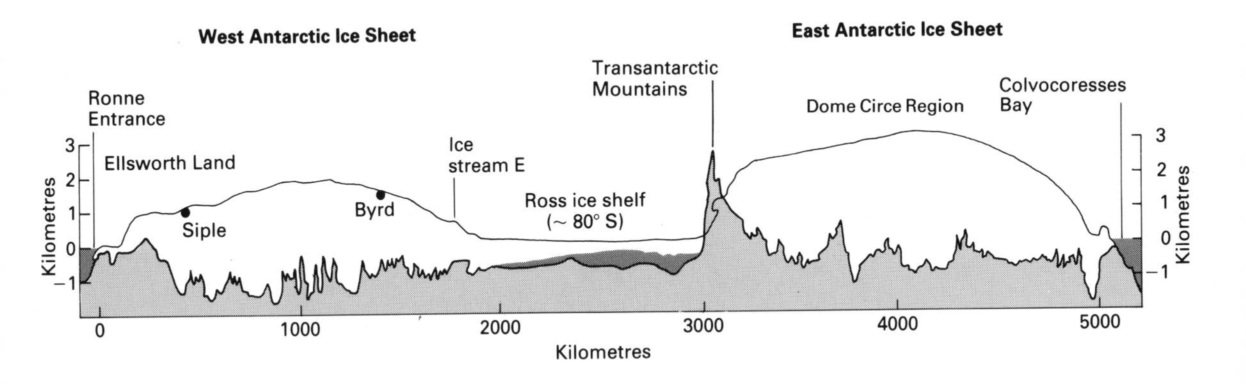

about twice the size of the UK and bigger than California. Periodically, large parts of these shelves break off as smooth, flat-topped, or 'tabular', icebergs, sometimes measuring more than a hundred kilometres across. Some ice shelves are stable over hundreds of years, but others can disintegrate rapidly. The Wordie Ice Shelf in the Antarctic Peninsula has largely disappeared in just 30 years, for example. Except for the Ward Hunt Ice Shelf on Ellesmere Island in the Canadian Arctic, which is insignificant in comparison, ice shelves are peculiar to the Antarctic.

Mountain glaciers

Less extensive than ice shelves, highland ice fields are still continuous over many square kilometres and bury many of the features of the underlying landscape. They are most common in polar and sub-polar regions, such as Spitsbergen, the Queen Elizabeth Islands in the Canadian Arctic, south-east Alaska and the Yukon, Patagonia and parts of the Antarctic Peninsula, although less extensive examples are found in many other high mountain areas in more temperate latitudes. The higher mountains below them project above the ice, with the ice surface between them undulating, the surface of highland ice fields thus

A high cirque on the 1916-metre-high peak of Oksskolten in northern Norway cradles the small glacier Skoltbreen. Below it flows the heavily crevassed valley glacier Austre Okstindbreen.

Above Mount Logan (6050 m), the highest mountain in Canada, towers above the upper reaches of Seward Glacier and dominates the vast highland icefield around it.

Right An extremely small cirque glacier near Titlis, a well-known mountain in the Swiss Alps, has the appropriate name 'Chli Gletscherli' ('small glacieret'). It is only 800 metres long, but it clearly shows the typical attributes of a real glacier: crevasses as a result of ice flow and a terminal moraine of debris deposited at the snout.

roughly reflecting the bed beneath. Valley glaciers commonly flow out in several directions from those ice fields located in temperate regions.

Many smaller glacier types occur in mountain regions. A common one is found at higher levels when ice erodes out and occupies hollows known as *cirques* (the French term for these forms, as in the French for 'circus'). Cirque glaciers are relatively short in relation to their width, mainly less than a few kilometres long, and some of them may spill out from their hollows.

High-altitude ice may flow downwards from cirques, ice caps, highland ice fields or ice sheets to form valley glaciers. These long tongues of ice flow downhill into regions well below the snow line, even occasionally into temperate rain forest, as in Alaska, New Zealand and Chile. They are typically tens of kilometres long and occur in many high mountain regions in the temperate and cooler latitudes. Some of the most dramatic examples descend into the Pacific Ocean from the highland ice fields of southern Alaska and the Yukon, where the Bering and Hubbard Glaciers (200 kilometres and 150 kilometres long respectively) are the longest in the Americas. In contrast, mainland Europe's longest valley glacier, the Grosser Aletschgletscher, is only 23 kilometres long.

In high latitudes many valley glaciers enter the sea, where they either remain grounded or float, and are then known as tidewater glaciers, although some glaciologists restrict this term to the grounded variety. Ice velocities tend to accelerate as the glacier enters water, producing exceedingly crevassed glacier tongues. The icebergs that are produced by calving are normally small and irregular in shape when they originate from grounded tidewater glaciers, but there are recorded instances of large tabular bergs from floating glaciers. Spectacular tidewater examples are found in the fjords of Spitsbergen, Greenland, Arctic Canada, Alaska, Patagonia, South Georgia and the Antarctic Peninsula.

Where mountain valleys open into larger valleys or on to plains, valley glaciers spread out into wide lobes called piedmont glaciers. The huge Malaspina Glacier in south-east Alaska, measuring 70 kilometres across, is the best known.

In some mountainous regions ice accumulates on slopes that seem to be almost too steep to hold any snow. These 'ice aprons' or 'hanging

Right Glaciers spilling out of their cirques, and descending towards the Glacier d'Otemma, one of the largest glaciers in Switzerland.

Below A typical floating tidewater glacier, the Gerard de Geer Gletscher, calving icebergs at the head of Kejser Franz Josef Fjord in central East Greenland.

Above Malaspina Glacier, Alaska, is the largest of the world's piedmont glaciers. It is well seen from an air route to Anchorage, and provides a superb example of folding due to glacier flow.

Left The 23-kilometre-long Grosser Aletschgletscher in Switzerland is a typical valley glacier of temperate mountain regions. Several smaller glaciers join to produce long narrow bands of debris, 'lateral moraines', in the main glacier tongue.

glaciers' are generally small, a few hundred metres across at most, and ice frequently breaks off them, making avalanches that are a particular hazard to anyone directly beneath. Large ice avalanches have sometimes wiped out whole villages in the Alps (as described in chapter 12). The ice debris from such avalanches, or indeed from the snouts of tributary valley or cirque glaciers that overhang the main valley, may accumulate sufficiently to produce a 'rejuvenated' or 'regenerated' glacier.

Warm and cold glaciers

A different way of classifying glaciers is according to their temperature distribution, in which respect there are two basic types. Temperate or warm glaciers are those in which the ice is at the melting point throughout, although a thin surface layer cools below zero Celsius in winter. Meltwater is abundant in summer and generally continues to flow to a small degree throughout the winter. Meltwater normally emerges through a portal in the middle of the snout, but beneath or within the whole glacier there usually is a well-developed drainage network, as well as a number of water pockets.

In much colder regions, where the mean annual air temperature is below zero Celsius, much of the ice is well below melting point. In the upper 12 metres or so the ice temperature fluctuates according to the season, but below that depth it is similar to the mean annual air temperature. Further towards the glacier bed heat flow from the bedrock warms the ice, perhaps to the melting point (as determined by pressure) at the bed if the ice is thick enough. This is true even in the heart of Antarctica where water pockets have been located beneath thousands of metres of ice, and where as much as a third of the ice sheet is wet-based. Where the bulk of the ice is below the appropriate melting point, the glacier is termed cold. When they occur in the Arctic such glaciers produce abundant meltwater during the brief summer, but the drainage pattern associated with them is different from that of temperate glaciers; stream channels only develop on the surface or develop close to or at the

Ice aprons or hanging glaciers cling to the precipitous face of Malte Brun (3155 m) above the Murchison Glacier in New Zealand's Southern Alps.

The Dry Valleys of Victoria Land are a unique polar desert, drier than the Sahara. Thermally, the glaciers are cold (with temperatures many degrees below melting point), and many are frozen to their beds. The two glaciers in this photograph of the Taylor Valley show contrasting behaviour. On the left the smoothly tapering front of the Crescent Glacier suggests slow retreat, while the Howard Glacier to the right has a vertical ice cliff that indicates advance. This late-winter picture illustrates the small amount of precipitation, with only small drifts of snow in the hollows of glacial debris.

glacier margins, as the streams which cut downwards cannot survive for long without freezing up.

The thermal characteristics of a glacier have a strong bearing on how they affect the landscape. In regions once covered by ice, meltwater channels and the distribution of stream deposits reflect which type of glacier occupied the area. Temperate glaciers and those parts of cold glaciers which slide on their bed erode the terrain strongly. Cold glaciers which are frozen to the bed are relatively passive. They erode very little and, in effect, protect the land from weathering and other kinds of erosion, such as by wind or by streams during the rare occasions that they flow.

Warm glaciers are characteristic of most mountain regions outside the Arctic and Antarctic; cold glaciers are the dominant type in the polar regions. But regardless of the classification used, the types described are merely convenient designations within a continuous spectrum of types of ice mass, and many glaciers are a combination of types. For example, a temperate glacier may have a cold section if it originated at a high altitude. Similarly, as we have already seen, many cold glaciers slide on their beds because a few metres of ice at the base has been warmed to the appropriate melting point by geothermal heat.

4

Fluctuating Glaciers

One of the commonest questions asked by passing walkers or climbers of glaciologists working on a glacier is, 'Are the glaciers advancing or retreating?' However, the answer invariably turns out to be less simple than is expected, because glaciers in the same area often exhibit differing behaviour.

The retreat of an ice margin is one symptom of deglacierization, albeit the most obvious, but it is not the only indicator. Some glaciers shrink principally by down-melting of the surface, but for the majority retreat of the margin is the more noticeable.

Adjacent glaciers can behave quite differently from each other. For example, on Axel Heiberg Island in the Canadian High Arctic, there are two large, cold valley glaciers, whose snouts are actually in contact with each other. Since observations began in 1959, the 14.5-kilometre-long White Glacier has been seen to be retreating consistently a few metres a year, as is reflected in the smooth rounded profile of its snout. In contrast, the 35-kilometre-long Thompson Glacier, which is a major outlet from the island's largest ice cap, has been advancing dramatically. Its spectacular, 30-metre-high, vertical cliff has been moving forwards at an average rate of 15 metres a year, sometimes collapsing and overriding the tundra plant life that has had several thousand years to become established. In addition, the Thompson Glacier is pushing a huge pile of river gravel forwards, like an enormous bulldozer.

Fluctuations in the Alps

Although the advance of the Thompson Glacier is exceptionally rapid for a cold, high Arctic glacier, it is eclipsed by much more dramatic fluctuations in temperate latitudes. Farmers in many glacierized mountain regions have carefully recorded the advance and retreat of glacier tongues, and particularly detailed records are available from the Alps, as meadows disappearing under advancing glaciers represent a serious loss to the local population. Over the centuries pastures above the timber line have been used for grazing cattle and sheep in summer, while many of the mountain passes have served as important merchant routes, so the memory of their loss has been passed down in legends for hundreds of years. A common version of these is that diabolical beings ate up the

Steep glaciers which receive a heavy snowfall, like the Franz Josef Glacier on the western side of the Southern Alps, are extremely fast-flowing and active, responding quickly to changes in the rate of accumulation. Glaciers such as this may flow down into regions where the temperature rarely falls below freezing and snow is infrequent. When this photograph was taken the glacier was advancing at a rate of over half a kilometre a year.

Close-up view of the 30-metre-high vertical cliff of the advancing Thompson Glacier on Axel Heiberg Island in July 1975. As the ice moves forwards at a rate of 18 metres a year, large blocks of ice fall from the cliff and are overridden. River gravels in the right background are being pushed forward into a series of ridges.

grasslands, as in the beautiful little Swiss village of Tiefenmatten, which was situated north of the Matterhorn, but which was buried by the advancing Z'Muttgletscher.

During the last few centuries tourists, especially painters, have documented the glaciers of the Alps. The Untere Grindelwaldgletscher in the Bernese Oberland has been painted so accurately and frequently that the history of its advance and retreat has been reconstructed with considerable precision as far back as 1600. The records show that from then until about 1860 this steep valley glacier advanced and retreated several times over a distance of about 600 metres. After 1860, the glacier began to retreat in an unprecedented manner, and except for some minor re-advances early this century, it lost more than 1800 metres of its length by the time it reached its historical minimum around 1977. The glacier then began a re-advance which by 1990 had amounted to more than 300 metres.

Most Alpine glaciers have broadly followed the trend of the Untere

Walkers approaching the glacier Vadret da Morteratsch in the Engadin area of eastern Switzerland cross terrain that shows various stages of recolonization by plants following the glacier's retreat. Posts mark past positions of the tongue and dramatically illustrate how far the glacier melted back in past decades. At the 1900-position of the glacier tongue (*top left*) larch trees have grown to well over 10 metres in height. At the 1920 position the larches are smaller and only a few pines are growing (*top right*). No coniferous trees can be seen at the 1960 marker in the photograph taken in 1986 (*above left*), but alder bushes, willow, grasses and flowers cover at least one third of the ground, thus preparing it for trees. Terrain which had emerged from under the ice only 16 years previously (1970-position) displays only small bushes but there are already dozens of species of Alpine flowers. Closest to the glacier (here shown in its 1980-position, *above right*) flowering plants had already become established on ground which had been exposed for at most three years.

Right The snout of Glacier d'Otemma in the Valais of Switzerland demonstrates the characteristic gently sloping form of a retreating temperate glacier. Medial and lateral moraines stand up as ridges because the debris protects the ice from ablation and slows down the retreat.

Grindelwaldgletscher. In 1890 systematic glacier measurements were initiated in Switzerland, and they have been continued to this day. The forest service workers and glaciologists who took on this considerable task of field work recorded the dramatic retreat of many Alpine glaciers. The Grosser Aletschgletscher, the largest glacier in the Alps, has lost 2.6 kilometres, or more than a tenth of its total length, since measurements began in 1860. Since the turn of the century, its average retreat has been 25 metres a year.

In front of other, more easily accessible glaciers in Switzerland, markers have been inserted to show visitors where the terminus was situated in former times. A particularly rewarding example is the walk from the railway station at Morteratsch (near the well-known resort of St Moritz) to the snout of the nearby glacier, Vadret da Morteratsch in the Engadin. Another example is the Rhonegletscher, the source of the River Rhône in the Valais, which can be reached by road. At both places, naturalists may observe the rate at which plants have recolonized terrain that had become ice-free only years or decades ago.

Left Heather Bay in Prince William Sound, Alaska, displays a fine push moraine formed by the Columbia Glacier in 1957. Taken in July 1989, this photograph shows how rapidly the glacier has retreated in recent times.

How did the climate change in order to cause such a dramatic glacier retreat? The temperature records from the northern Swiss city of Basel, which began in 1755, show that cold summers were much more common during the period from the mid-eighteenth to the mid-nineteenth century, coinciding with the phase of enhanced glacier activity known as the Little Ice Age. Summer temperatures affect glacier melting, and are therefore much more significant for a glacier's mass balance than are winter temperatures. At the same time precipitation did not decrease. Thus temperature seems to have been the major controlling climatic factor. Even so, the size of these temperature changes was surprisingly small (one or two degrees Celsius of summer cooling) considering that the retreat of a typical glacier tongue has subsequently been around a kilometre. We see from this that glaciers in temperate regions are extremely sensitive to temperature changes, and can therefore be used to reconstruct climatic trends in areas with poor or non-existent climatic data.

The 1950s brought fears that many Alpine glaciers might disappear altogether, following the long period of slowly rising temperatures since the Little Ice Age, especially with some record highs in the 1940s. However, a slight cooling then set in, which lasted into the 1970s, and increasing numbers of glaciers began to advance. By the late 1970s more of the regularly documented Swiss glaciers were advancing than were retreating, at least making the glaciologists happy, if no-one else. Some Swiss glaciers responded with positive gusto to the new cooling phase; for example, the Allalingletscher advanced a record 174 metres in 1970–71 and the Oberer Grindelwaldgletscher covered 100 metres in 1971–72. However, the response time of some of the larger Alpine glaciers has been too slow to take advantage of this cool phase, which in any case only proved to be short-lived: the reaction time of the Grosser Aletschgletscher is so long that it failed to stop retreating before the warming phase of the 1980s took hold. The temperatures have been so high in recent years that, at the time of writing, the majority of Alpine glaciers are again on the retreat.

The most easily visited part of a glacier for walkers is usually the snout, where an enthusiastic observer might, over several years, enjoy undertaking a simple but rewarding photographic project. Photographs taken from the same spot, especially of a side view, will clearly document a glacier's retreat or advance over a year. However, even by visiting a glacier only once it is usually possible to judge whether it is advancing or retreating.

Retreating glaciers have a gently graded, flat snout, so that it may be fairly easy to walk on to the ice. Large areas of stagnating debris-covered ice will give rise to unstable and hummocky topography. These may well be a meltwater stream emerging from a wide-open ice cave or glacier portal. And there will be no plants on the ground adjacent to the ice since it will have only recently been uncovered.

An advancing glacier snout, in contrast, usually has a steep, convex front which may be difficult to scale except by experienced ice climbers. The meltwater stream may emerge from the ice without creating a cave, since the ice is trying to squeeze shut any opening. During the short advance phase of the late 1970s, several Alpine glaciers reached terrain that had been ice-free for several decades, so that trees had had time to establish themselves. Seeing trees that are being pushed over by the advancing Ghiacciao del Belvedere has become a dramatic illustration of glacial advance near the village of Macugnaga in the Alps.

Many glaciers show annual advance-retreat cycles. Even glaciers whose ice margins are in overall retreat may advance briefly in winter when there is little melting, creating a sequential set of small annual push ridges or moraines.

Tidewater glaciers

Glaciers ending in the sea are especially prone to rapid retreat. The most dramatic developments have been along the south coast of Alaska, where glaciologists have noticed that the deeper the water is at the glacier terminus, the more intensively it calves and produces icebergs. Reduced ice supply from the accumulation area has led to the retreat of several very long valley glaciers that terminate in fjords. Fjords are commonly deepest in their inner reaches because erosion by ice has been at its greatest there, so that as the glaciers recede into deeper water and contact with the bed is weakened, their retreat becomes more pronounced.

Glacier Bay, a fjord system up to 550 metres deep in the Panhandle of Alaska, displays a remarkable record of glacial retreat over the last two

Left Glacier Bay in Alaska's Panhandle reveals a dramatic example of glacier retreat. Since 1794, when the bay was filled by ice, the glaciers have retreated 135 kilometres, leaving a complex network of deep fjords and islands. As the land is first released from its burden of ice, the ground is characterized by bare rock, debris and dead ice. The photograph far left illustrates the calving snout of McBride Glacier in June 1986 as it retreats away from Muir Inlet. Within a few years, small plants such as fireweed (*left*) become established. Then within a decade impenetrable alder has developed. In turn this is replaced by willow and cottonwood. The photograph below left shows Glacier Bay from a position on the east side that was covered by ice only a century ago. Apart from the wide expanse of open water between here and Mount Fairweather (4670 m) in the background, dense mixed forest of deciduous trees and early spruce colonizers fringes the water's edge, providing a home for many species of animal. Finally, at Bartlett Cove at the mouth of the bay, there is now a mature spruce forest among the boulders of the 1750 moraine.

centuries, combined with rapid colonization by a succession of different plant species. In 1794, near the end of the Little Ice Age, the British Navigator George Vancouver observed at the mouth of what is now Glacier Bay 'an immense body of compact perpendicular ice, extending from shore to shore, and connected with a range of lofty mountains on each side'. By the end of the nineteenth century the ice had retreated over 60 kilometres up the bay, but at the same time had filled it with many icebergs. Today, the head of the fjord is 135 kilometres from the position at which Vancouver observed his ice wall at the mouth of the bay, and icebergs are few in number, except close to the existing tidewater glaciers.

Most glaciers in Glacier Bay continue to recede at a rapid rate, but

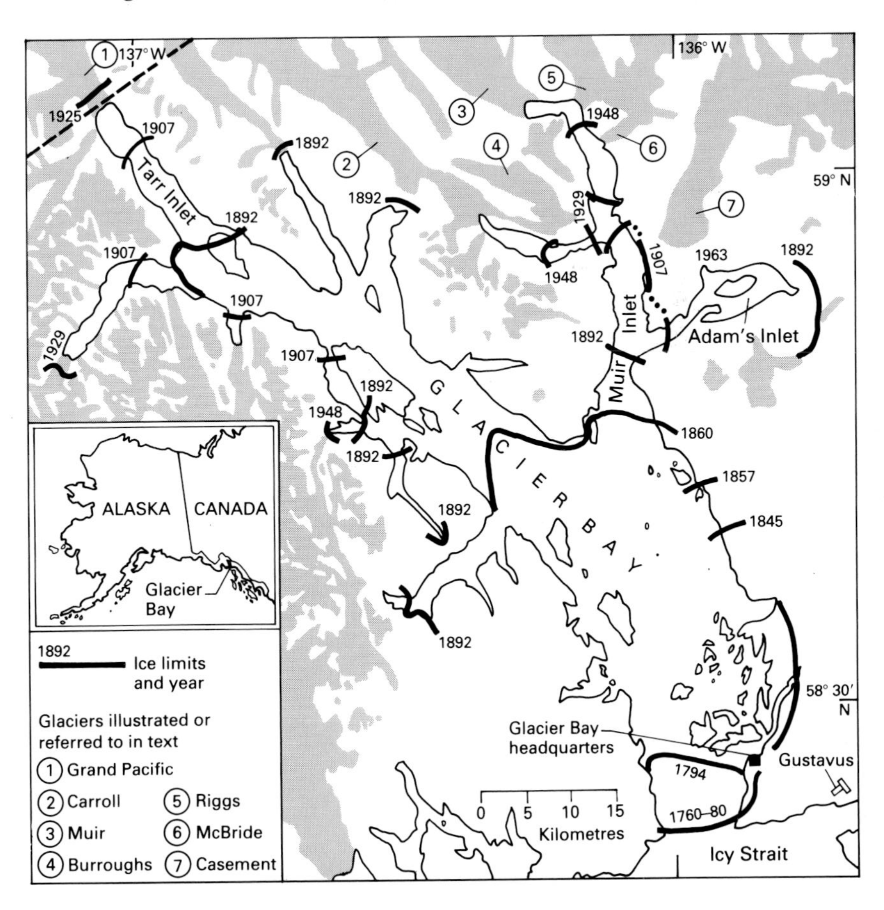

Right Map of Glacier Bay, Alaska, showing the present distribution of glacier ice and the former positions of the glacier margins in the Bay.

some fed by the higher mountains of the Fairweather Range and the St Elias Mountains are now advancing. One glacier, the Grand Pacific, caught in this ebb and flow is unsure about its national identity! By the 1920s it had retreated so far that marine waters extended into Canadian territory, separating parts of Alaska's Panhandle from the rest of the state, but a slight re-advance has since brought the terminus back into the USA.

The phenomenal retreat of the glaciers in this area has been accompanied by some of the most active geological processes on earth. Partly owing to collision between the Pacific Ocean plate and the North American continental plate, and partly to upward readjustment of the land following removal of the ice burden, there is a rapid crustal uplift in this earthquake-prone region. At Bartlett Cove, near the mouth of the bay, the land is rising at a rate of four centimetres a year, having risen several metres since the 1790s. A combination of unstable land, loose glacial debris and high precipitation also creates rapid erosion and deposition, and some of the newly vacated fjords are filling with sediment at a rate of several metres a year.

Colonization by plants is equally rapid, and each succession makes a contribution to the building of the sitka spruce forest now developed on the Little Ice Age moraines. As the ice recedes from the land, the first plants to appear are lichens and mosses. Mat-forming Dryas and other low-growing plants follow. These give way in turn to alder, willow and cottonwood, and finally to sitka spruce. The climax of the succession has yet to be reached in Glacier Bay, but a few western hemlocks are beginning to make their appearance in the region of Bartlett Cove. The inner branches of the bay recently vacated by ice provide evidence that coniferous forest grew all over the region prior to the Little Ice Age. Trees pushed over thousands of years ago by advancing ice have been preserved in glacial sediments and are now emerging in areas prone to erosion.

Icy Bay, to the north-west of Glacier Bay, is another fjord system to have emerged in recent historical times. Most of the retreat is recent in origin having taken place this century; the movement is most vividly seen in the current accelerated retreat of Columbia Glacier in Prince William Sound.

The contribution of man to glacier retreat

Glaciers have been induced to retreat by creating lakes and allowing the glaciers to calve into them. Dams for hydro-electric power generation were constructed in front of Ghiacciao del Sabbione and Griesgletscher, two Alpine glaciers which have adjacent accumulation areas on the Swiss–Italian border, and the filling of the resulting lakes led to a rapid retreat of the ice. This had been taken into account when planning the storage capacity of the reservoirs. A similar, but rather bigger project is now being evaluated in the Bernese Oberland of Switzerland, where a large dam is planned that would flood Unteraargletscher and shorten it by three or four kilometres. At the time of writing it is not clear whether environmentalists might be able to stop the project which, apart from destroying a most interesting and historically important glacier tongue, would alter an area of outstanding natural beauty beyond recognition.

Man-induced glacier retreat. The creation of a reservoir for hydro-electric power has led to the rapid retreat of Griesgletscher, Switzerland, and the development of an ice cliff.

5

Ice on the Move

As well as advancing and retreating, glaciers actually flow. The curving of crevasses and banded structures known as ogives, the recordable movement of surface rocks over a few days, and the occasional cracking and creaking sounds within the ice are all symptoms of this. Ice flow is also indicated by the eroded rocks and deposits that are left behind after the ice margins have retreated. The nature of the ice flow, in fact, very much determines the character of a glacier.

Rates of movement of flowing glaciers are extremely variable. Some small glaciers and ice caps may flow only a few metres a year, or are motionless. The fastest parts of most reasonably sized valley glaciers flow anything between 50 and 400 metres a year, even faster if they end in the sea, while large ice streams in Antarctica and outlet glaciers in Greenland flow steadily at a thousand metres a year or more.

A small percentage of glaciers flow in a rather unpredictable manner; they remain relatively inactive for many years, but may accelerate suddenly, even increasing many hundredfold in speed. For a period of a few months they advance over distances measured in kilometres, in what are known as surges, phenomena that have given rise to the colloquial term 'galloping glaciers'.

The flow of ice

Ice flows by two main methods: by internal deformation and by sliding on its bed.

Internal deformation

As snow turns to firn and then ice, its constituent crystals alter under the weight of material accumulating above them and under the influence of gravity. These stresses cause the ice to change shape in a rather plastic manner, much as soft putty or porridge will deform on a slope, only considerably more slowly. Deformation is greatest near the bed and sides of a glacier, so a typical flow pattern shows an initial rapid increase in velocity away from the margins, then a declining rate of increase towards the middle.

Such a velocity profile is parabolic in shape. Similarly, there is a rapid increase in velocity in the first several metres above the bed of a glacier, then the rate of flow remains essentially constant through the rest.

Previous page The Mer der Glace, the longest glacier in France (12 km) descends from Mont Blanc (4807m) at upper right. The faster flow of ice in the centre of the glacier is indicated by the light and dark curved stripes ('ogives') which form in the icefall.

Icicles at the side of a steep mountain glacier in the Cordillera Blanca of Peru illustrate the basal movement of the ice. These icicles formed in the evening when meltwater flowed; later, during the cold of the night, the meltwater stopped. Icicles which had reached the firm ground have bent as a result of the glacier's motion towards the lower right.

Glacier flow has a profound effect on the nature of the ice. The upper layers, down to depths of 30 metres or so in temperate glaciers, and more in cold glaciers, are brittle under tension. When they move they crack, creating one of the most hazardous features of glaciers – the open fractures called crevasses. Less dramatic, but equally characteristic, a wide range of other structures form, reflecting deeper plastic deformation over a long period, such as folds and foliation.

Basal sliding

The second component of glacier flow is basal sliding, whereby the glacier slips over its bed. Large quantities of meltwater produced in summer reduce the friction between a glacier and its bed and cause faster flow. In a temperate glacier, basal sliding is the major component of its

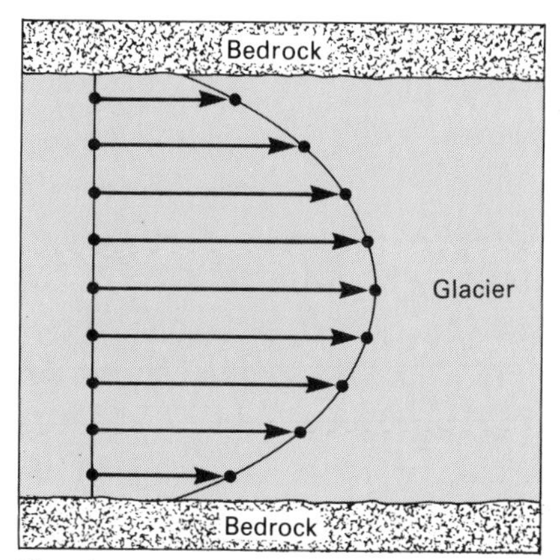

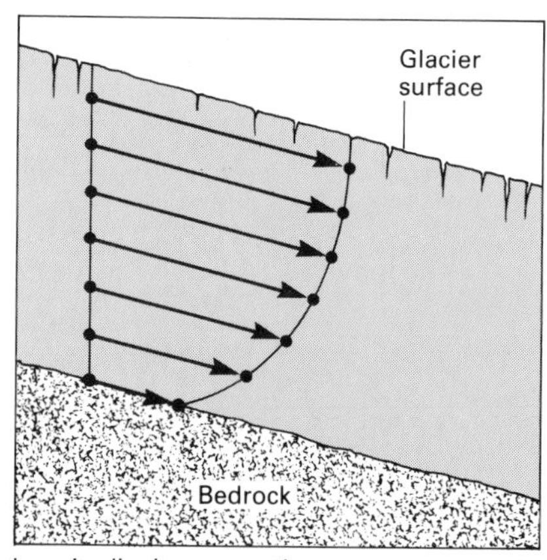

Ice flows by internal deformation. The arrows depict the distance particles will travel over a period of several years. Those in the centre and in the upper part of the glacier travel the furthest.

Some examples of ice velocities in different types of glacier

Glacier	Region	Centreline velocity (m/yr)	Comments
Lambert Glacier	East Antarctica	347	Part of largest glacier drainage system in Antarctica
Amery Ice Shelf	East Antarctica	1200	As above
Byrd Glacier	East Antarctica	760	Glacier drainage from polar plateau through the Transantarctic Mountains
Ice in West Dronning Maud Land	East Antarctica	1 to 15	Slow moving part of ice sheet where unchannellized
Jakobshavn Isbrae	NW Greenland	4700	World's fastest outlet glacier
White Glacier	Axel Heiberg Island, Canada	40	Cold, sliding valley glacier
Grosser Aletschgletscher	Swiss Alps	200	Fastest part of largest glacier in the Alps
Griesgletscher	Swiss Alps	40	Small valley glacier
Saskatchewan Glacier	Canadian Rockies	117	Valley glacier
Charles Rabots Bre	Okstindan, Norway	8	Steep, thin cirque glacier

Contribution of basal sliding to total glacier flow from tunnel and borehole measurements

Glacier	Region	Basal sliding %	Ice thickness (metres)
Grosser Aletschgletscher	Swiss Alps	50	137
Tuyuksu Glacier	USSR	65	52
Salmon Glacier	Canadian Rockies	45	495
Athabasca Glacier	Canadian Rockies	75	322
Saskatchewan Glacier	Canadian Rockies	20	555
Blue Glacier	Olympic Mts., Washington	90	26
Vesl Skautbreen	Norway	9	50
Meserve	Victoria Land, Antarctica	0	80

flow, and may account for as much as 90 per cent of its total overall movement.

Where basal sliding occurs over uneven bedrock it often generates caves where glaciologists have been able to observe at first hand the range of processes of erosion and deposition – albeit in uncomfortable and not always safe circumstances!

Since sliding velocities are related to the amount of meltwater available, glaciers move faster in summer than in winter, and faster in daytime than at night. Exceptional speeds may also be induced by heavy rain. In cold glaciers basal sliding can only occur where the ice is thick enough for the base to be warmed to melting point by geothermal heat. Moreover, it is quite common for the snouts of otherwise sliding glaciers to be frozen to the bed because the ice is thinner there and thus affected more by the low mean annual air temperature.

There is a third factor affecting the rate of glacier movement as well as internal deformation and meltwater at the bed; heavy accumulation of snow creates dynamically more active glaciers that flow at rates of up to several hundred metres a year. With all glaciers, as might be expected, steep ones flow faster than gently graded ones, and as a result are generally thinner.

Movement over soft, deformable beds

A layer of unconsolidated sediment frequently underlies moving ice, known as till. This is a mixture of particles of all sizes from clay to boulders. 'Till' is an old Scottish term for stiff, indurated stony ground, but is now applied internationally to deposits released directly from ice. The English term 'boulder clay' is well established but is no longer favoured, since not all tills contain either clay or boulders. When saturated with water, this sediment deforms more easily than the basal ice, and glacier movement is assisted by shearing within the soft, deformable sediments, rather than by sliding.

Structures in glacier ice

A glacier can be considered as a small-scale model for the processes that occur deep in the earth's crust, such as the formation of the Western

Cordillera or the Alps as a result of the collision of continental plates. If you walk over bare glacier ice or peer down into crevasses you can see a wide range of layered structures. Like rocks in mountain ranges, these may be continuous or discontinuous, are often folded, and can form complex patterns. All result from glacier flow, and mostly they reflect the plastic behaviour of ice deep in the glacier. On the smooth surface of a glacier after rain they appear as beautiful layers of contrasting colour and texture, from blue to white, and coarse-grained to fine-grained.

The accumulation of snow

If we look at a glacier we can determine how the different layered structures evolve from the initial snowfalls in the accumulation area. After the year-by-year accumulation of snow and its subsequent change into

Deep in a glacier the ice is less brittle than near the surface. Instead it is rather plastic, allowing the development of fold structures. In this example from Gornergletscher, Valais, Switzerland, the original annual layers of firn have become folded. As the ice ablates the deeper ice that has flowed in a plastic manner becomes exposed. (An ice axe indicates scale.)

Left Foliation, when weathered, is often quite pronounced on a glacier surface, the darker layers melting the most and trapping dust. This photograph shows longitudinal foliation running the length of the tongue of the Orange Glacier in south-central Alaska. The Hubbard Glacier is in the background.

Below left Continued flow of already folded ice may refold existing folds, creating 'eyed' folds, so called because they may resemble eyes in animals. This August 1975 photograph shows the advancing frontal cliff of Crusoe Glacier on Axel Heiberg Island, in the Canadian Arctic. The 'eye' near the base represents an overturned fold that has been compressed parallel to the ice front. Horizontal foliation forms the bulk of the ice in the cliff above.

Below Flying along the south coast of Alaska, airline passengers obtain a stunning view of enormous folds in the ice of the Bering Glacier.

firn, then ice, a layered structure called sedimentary stratification develops. Individual layers consists of thicker parts of light blue, coarse-grained bubbly ice up to a few metres thick, separated by thin layers of dark blue, coarse-grained clear ice. The former is the result of direct conversion of snow to ice by pressure as further snow accumulates above, whereas the latter reflects local horizontal zones or layers, which had become saturated with water during the melt season and subsequently refrozen.

As the ice moves downhill the layers deform gently, because of the plasticity of the ice and the faster movement at the middle of the glacier than at the margins. A period of excessive ablation may remove and truncate many of the layers, so that when new layers accumulate a marked discontinuity, called an unconformity, is apparent.

Foliation

As flow continues, the layers become more deeply buried and plastic deformation results in the stratification becoming more and more tightly folded and sheared on both a large and small scale, giving rise to a new layered structure called foliation. Both folds and foliation generally develop at depth in the plastic flow zone.

Foliation consists of the same sorts of layers as stratification, but the layers are closer together and less continuous. The structure is most strongly developed where shearing is greatest, such as close to the glacier margin, or where two ice streams combine. In such cases the coarse ice crystals may become unstable and break down into fine-grained ice which is of whitish, granular appearance. Normally, foliation of this type is parallel to the ice margin, which in thick mist gives an observer a useful indication of the way up or down a large glacier.

As ice moves downslope and one set of structures has formed, conditions may be ripe for a new set to develop. Thus a new generation of foliation may evolve by folding of an earlier foliation in a similar way to folding in stratification higher up the glacier. This process may be repeated several times. Thus, ice in the lower reaches of a glacier may possess an exceedingly complicated structural pattern. Although all these structures form at depth, they are ultimately exposed at the surface by ablation.

Both these two glaciers in the Italian Alps, Vedretta di Fellario (left) and Vedretta di Fellario Orientale (right) have unusually closely spaced ogives, which have developed below short, steep icefalls. The peaks at the head of these glaciers include Piz Bernina (4020 m; left) and Piz Palü (3905 m; right)

Veins related to fracturing

Other sorts of layers result from the fracturing and stretching of ice, but in addition to those that arise from the generation of open crevasses. Near-vertical faults form by displacements along fractures both vertically and laterally. Low-angle faults called thrusts may generate at the

glacier bed and extend upwards and forwards, such as where the ice is slowing down. Such features are also accompanied by folding in many cases, and are especially prone to develop in cold glaciers where the conditions at the bed change from basal sliding to frozen.

Many crevasse sets are associated with veins of clear blue ice which are either the traces of old, water-filled crevasses, or are formed as a result of recrystallization of ice in narrow zones under tension. These features, known as crevasse traces, develop with continued flow into a different sort of foliation from that described above. They are most commonly found as transverse sets which develop into a curved structure known as arcuate foliation. When foliation forms longitudinally the layers are less continuous. However, the two types may merge into one another because the faster flow in the middle of the glacier produces crevasse traces with longitudinal orientation near the margins.

Among the most striking of all glacier structures are ogives – curving bands which form within and below the chaos of collapsing blocks in an ice-fall. They are made up of sets of light and dark bands or waves, each usually several metres wide. These bands occur in pairs which curve across the whole of a glacier in regular fashion. They only develop within ice-falls, but for some reason not all ice-falls generate them.

In detail, the individual light and dark bands each comprise many layers similar to arcuate foliation, and probably stem from transverse crevassing. Ogives are annual features, a pair of light and dark bands or a wave representing a year's movement through the ice-fall. Thus, by measuring the distance across a pair of light and dark layers we can easily determine the approximate ice velocity.

Banded ogives, also known as Forbes' bands after the Scottish geologist who first described them, reflect the passage in alternate seasons of dirty ice in summer and snow-covered ice in winter through the ice-fall, hence the light and dark bands. Wave ogives reflect the passage of thinner ice (because of ablation) through the ice-fall in summer. Both types are surprisingly persistent, and can often be traced all the way to the snout, as on the Mer de Glace in the French Alps, where the best-known examples occur. Distance between the bands is not the only guide to glacier speed: by counting the number of ogives we can determine how long it has taken ice to cover the distance from the ice-fall to the snout, as in the Mer de Glace which has about fifty. Also as ogives move down glacier they often become more closely spaced, indicating that the ice is slowing down.

As all these structures, except ogives, have direct counterparts in rocks in mountain belts, it is common to see structurally complex rocks

Many fractures or faults do not open into crevasses. In this photograph a series of closely spaced faults has cut through the debris-rich layers at the base of Taylor Glacier in Victoria Land, Antarctica. Ice flow is to the right.

Right Crevasses in the ablation area of a glacier quickly melt back to form less steep sides, while meltwater gathers in their bottoms. Matanuska Glacier, Alaska.

alongside glacier ice that deformed in a similar way, though at a much slower rate and at a much higher temperature.

Crevasses

Crevasses – great V-shaped clefts many times deeper than they are wide – are among the most serious hazards faced by the mountaineer. Many of them are covered by bridges of snow, which hide them from unwary travellers who can fall to their deaths when they unwittingly try to walk across them. In good light you can tell a bridged crevasse from the subtle shades of blue on the snow's surface, but in cloudy conditions there may be no visible sign that the area is crevassed. Thus, you have constantly to probe the snow with an ice axe if you are to be safe.

New crevasses are generally clean-cut and straight-sided. Those in the accumulation area are more dangerous than the ones in the ablation area after the winter snow has melted, as they are less easy to see. As they get older and melt on their passage downwards the sides of the crevasses become less steep and more rounded, and in this state you can walk into

Scientists are here exploring a clean-cut crevasse typical of the accumulation area of a glacier, on Charles Rabots Bre, northern Norway. The internal structure of the glacier can be seen in the vertical walls: the steep layers behind the man on the rope ladder represent annual accumulations of firn (stratification) that has been tilted by glacier flow. Layers of shallower inclination above the man's head are also of stratification. The discontinuity between the two sets represents an unconformity, created as a result of prolonged erosion of the underlying layers.

them without difficulty, although many contain lakes. Eventually, melting will proceed until little trace remains.

Rescue from freshly formed crevasses in the accumulation area may be difficult, since they have overhanging sides, and considerable expertise in handling ropes is essential for anybody crossing fields of snow-covered crevasses. It is unlikely, however, that any unlucky faller would actually go right to the bottom of one, because they usually contain old collapsed snow bridges that will break a fall. As the ice moves, crevasses close and there have been instances in the Alps of unrecovered bodies becoming embalmed in the ice and only released after many decades. One well-known case took place in 1820 when three guides in a climbing party were swept by an avalanche into a deep crevasse on the Glacier des Bossons on the slopes of Mont Blanc near Chamonix in the French Alps. It was only after forty-three years that the bodies were released from the ice near the glacier's snout, having travelled just over 3 kilometres.

The mystique of crevasses and their frequent appearance of bottomlessness has led to many exaggerated ideas about their depths. Claims of instances hundreds of metres deep are not unusual, yet in the temperate glaciers so typical of alpine regions they are rarely open to depths of more than 30 metres. On the other hand, the cold ice sheet of Antarctica does have huge ones that could easily swallow objects as big as houses, let alone over-snow vehicles. However, few reliable measurements of their depth exist.

The principal reason why crevasses are not 'bottomless' is that only the upper part of the glacier is brittle. Below a critical depth (around 30 metres in temperate glaciers) the weight of overlying snow and ice makes the ice more plastic, so any fracture that spreads from the surface cannot open up below that depth, as the ice will flow faster into a space than it can split. The exceptions to this rule occur if there are already weaknesses in the ice, say from older fractures formed further towards the head of the glacier, or if the crevasse is filled with water.

Types of crevasse

Crevasses form where the ice is under tension, such as where a glacier flows over a step in the bedrock, round a bend, or where the valley that

As an ice-fall steepens and the degree of transverse crevassing intensifies, the ice breaks into irregular blocks called *séracs*. These unstable features collapse frequently, making travel through icefalls difficult and hazardous. This view is of the top of the icefall in the Glacier de Saleina in the Valais, Switzerland.

constrains it narrows or widens. They normally form in well-defined sets, classified according to their geometry: longitudinal, marginal, transverse, splaying and *en echelon*. Sometimes, however, several sets of crevasses may intersect, creating a chaotic, totally broken-up surface, with ice towers known as *séracs*. Where a glacier flows over a pronounced step, the surface first fractures into transverse crevasses, before breaking up totally, creating chaotic reaches known as ice-falls. Séracs are very unstable, and where possible ice-falls are best avoided because of the danger of their collapsing.

A special type of crevasse, known as a *bergschrund*, occurs at the head

Plan view of the principal types of crevasse in the tongue of a valley glacier, together with the types of flow involved. The arrows, which are at right angles to the crevasses, indicate the directions in which the ice is pulled apart.

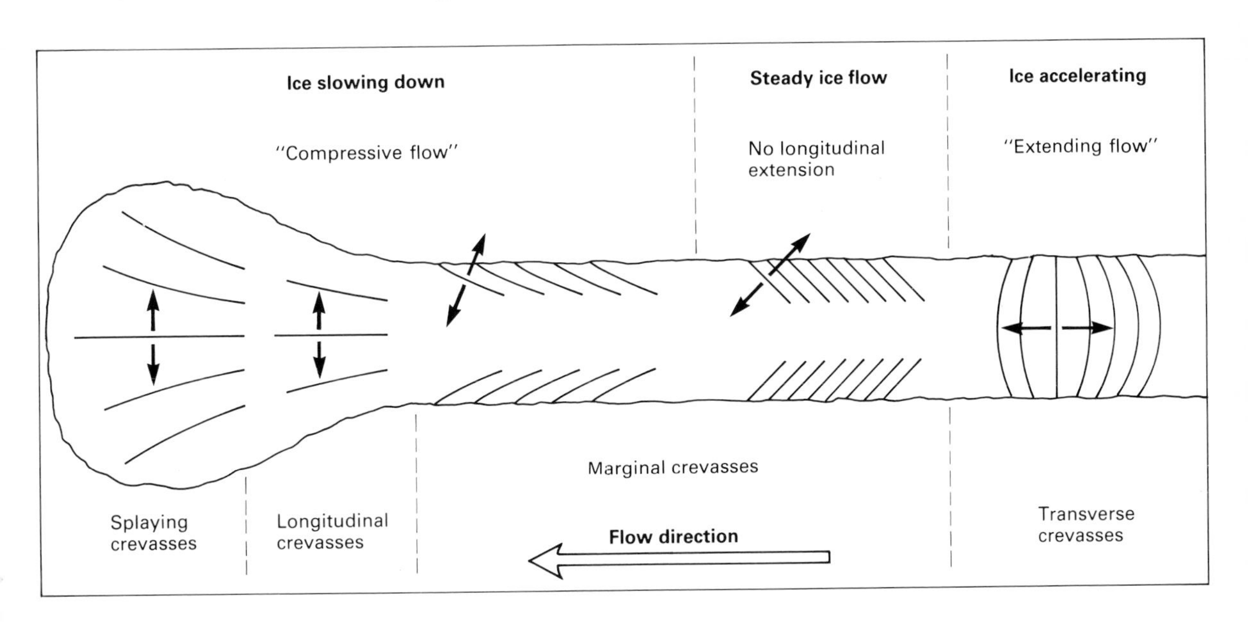

Left The Glacier d'Argentière descends from the north slopes of the Aiguille Verte in the Mont Blanc range of France to moderate altitudes where an abundance of alpine flowers blooms in July. The long and relatively narrow Glacier d'Argentière is a typical valley glacier, and has a prominent icefall in which the ice surface is fractured into small blocks.

Below Intensive crevassing leads to the total break-up of the blocks between the crevasses, leaving séracs, such as these in Steingletscher, Graubünden, Switzerland.

The 3000-metre-high Main Divide in the Southern Alps of New Zealand, just north east of Mount Tasman, receives huge quantities of snow, and the glaciers on the precipitous eastern slopes are extremely active and heavily crevassed. The higher irregular crevasses on such slopes are known as *Bergschrunds*.

of a glacier, generally where the gradient is very steep and the main body of ice pulls away from the more stable ice that adheres to the steep mountainside. They are irregular and individually may extend laterally for many hundreds of metres. Unless they are bridged by snow, they form a difficult barrier for mountaineers to cross.

Sometimes there is another cleft in the glacier higher still, adjacent to the rock wall; this is known as a *randkluft*. Such clefts form where a rock wall absorbs radiation and melts the adjacent ice.

Surging glaciers

In what is a rare but spectacular type of ice movement, in the spring of 1986 one of the longest glaciers in North America, the 150 kilometre-long Hubbard Glacier in southern Alaska, accelerated dramatically. The ice advanced at around 10 metres a day at the snout, practically ten times its normal speed. This rapid advance was said by some glaciologists to have been triggered by a surge of its tributary, the Valerie Glacier, which at the time reached speeds in excess of 40 metres a day, many times in excess of its normal rate. A tongue of the Hubbard Glacier advanced across relatively shallow water at the mouth of Russell Fiord, over a small forested island, smashing trees as it went. In just two months it had reached the mountain wall on the far side of the fjord, creating a huge lake 50 kilometres long out of the fjord.

Many marine mammals, including hundreds of seals and porpoises, became trapped in water that became less and less salty and increasingly murky from glacial sediment. Environmentalists managed to save some of them by carrying them around the ice barrier and releasing them into the open sea, but many had to be left. As the lake rose further with water from the glacier and from melting snow, numerous birds were driven from their nests, and many eggs and chicks were destroyed. By early October Russell Lake, as it was by then known, had risen to nearly 30 metres above sea level and had begun to threaten the village of Yakutat and the airport that lay just beyond the head of the lake.

Fortunately for Yakutat, the lake broke through the ice dam on 8 October, but by then many birds and mammals had drowned and marine life had perished in the increasingly fresh water. At its peak, discharge from the lake following the outburst was estimated as thirty-five times the flow over the Niagara Falls. Stationary waves of water reached nearly 10 metres in height, and huge amounts of ice fell into the river and on into Disenchantment Bay.

Standing on the shore of this lake in July 1986 was an unforgettable experience. The jagged wall of ice of Hubbard Glacier constantly sent large masses of ice crashing into the lake, the initial fracture sounding like a thunder crack and the break-up a muffled explosion. Small icebergs littered the lake surface in front of us and bobbed gently among the drowning trees. Agitated birds fluttered frantically back and forth, seeking their flooded nests. All the while the water crept gradually up the flat plain on which we were standing.

From a high vantage point, the glacier surface was an amazing sight: totally broken up into huge crevasses as far as the eye could see. The ice pinnacles, perched precariously between the crevasses, would occasionally topple and break up into a pile of rubble, a sharp disturbance in the still, clear air.

Whether or not the Hubbard Glacier advanced rapidly in response to a surge of its tributary, many glaciers in the area do undergo short-lived phases of accelerated flow, between quiescent phases lasting a few years to decades. For example, in 1982–83, after eighteen years of relative inactivity, the adjacent, Variegated Glacier surged. In this case glaciologists predicted the surge, since four events of this nature had been recorded earlier in the century, and they were able to investigate the event and its aftermath, with the result that the Variegated Glacier has yielded more information about surging than anv other glacier. During the surge, the surface of the glacier broke up totally into a maze of crevasses and chasms as the ice reached top speeds of 65 metres a day. Ice in mid-glacier was displaced by about 2 kilometres during the eighteen-month-long surge, compared with only 1 kilometre in the entire period of seventeen years preceding it. By 1986 the crevasses had melted back so that, although there remained a chaotic maze of hills and valleys of ice, it was possible by wearing crampons and taking a great deal of care and effort to walk around and examine the structures that had resulted from the surge.

The distribution of surging glaciers

The geographical distribution of surging glaciers is peculiar and does not follow any obvious laws. Scandinavia, the Southern Alps of New Zealand and the European Alps do not have surge-type glaciers at present (although one glacier in Austria is thought to have surged in the past). In Alaska and the Yukon surge-type glaciers are confined to the St Elias Mountains, the Alaska Range, the Wrangell Mountains and the Chugach Mountains. Others occur in the Arctic Queen Elizabeth Islands. However, there are none in the more southerly parts of the Rocky Mountains of Canada and the USA. Elsewhere surge-type glaciers are found in the Andes, Iceland, Greenland, Svalbard, several parts of the USSR and the Karakorum. Thus they occur in a wide range of topographic and climatic setings, but they form no more than a small percentage of the total number of glaciers in any one area.

Some glaciers, such as the Jacobshavn Isbrae of north-western Greenland, flow constantly at 'surging' speeds, and others, like the Fox and Franz Josef glaciers of New Zealand occasionally advance at very fast rates. However, none of these are strictly surge-type glaciers because they lack a dormant phase.

One of the great unresolved questions is 'Does the Antarctic Ice Sheet surge?' Some glaciologists have suggested that one of the largest ice drainage basins of the ice sheet, the Lambert Glacier-Amery Ice Shelf system, might currently be building up to a surge. Others disagree, and the data available is inadequate to come to any definitive answer. The West Antarctic Ice Sheet is considered by some glaciologists to be particularly unstable because it is grounded below sea level and is held back

The rapid advance of Hubbard Glacier created much public interest as it dammed Russell Fiord, threatening the village of Yakutat and its airport, and causing the death of many animals. These two photographs were taken from similar positions, prior to the surge (in 1983) and during the surge in July 1986. In the photograph above, Disenchantment Bay, in the background and extending to the left, and Russell Fiord in the foreground are connected. Hubbard Glacier is entering the water from the right, with the separate Turner Glacier and spurs of Mount Cook in the background. In the photograph on the next page a tongue of ice from Hubbard Glacier has blocked off the mouth of Russell Fiord creating a huge lake. The rate of advance of the ice at this time was around 10 metres a day. By October 1986 the lake level had risen 30 metres, but then the dam broke.

by ice shelves which may rapidly disintegrate as the climate and bordering ocean warms. However, there is no evidence yet that the huge southernmost ice shelves, the Ross and Filchner-Ronne, are disintegrating, although the smaller ones further north in the Antarctic Peninsula are doing so. A major surge of either the Lambert Glacier or the West Antarctic Ice Sheet into the Southern Ocean would have a drastic effect on the world's weather, and as the ice broke up, huge volumes of water would be released, raising the sea level globally by several metres. And if even larger parts of the Antarctic Ice Sheet were to surge, then low-lying countries such as the Netherlands and Bangladesh, not to mention many large cities, would become flooded. This may not be as far-fetched as it seems. There is geological evidence that the Laurentide Ice Sheet of North America surged repeatedly over the Mid-West during the last (Wisconsinan) Glaciation, about 10,800 to 14,000 years ago. Similarly, some geologists have proposed that an ice sheet over the Barents Shelf surged over northernmost Europe during the late Weichselian Glaciation at about the same time.

The nature of surges

Some surges occur at regular intervals. For instance, the 1982–83 surge

of Variegated Glacier was predicted on a known periodicity of approximately nineteen years. The Medvezhiy Glacier in the Soviet Pamirs surges about every ten years. However, the surge cycle for most glaciers is much longer and therefore difficult to predict. Between surges glaciers experience years or decades of inactivity, with melting of stagnant or slow moving ice, and the frequent development of a complex internal drainage network, including numerous large potholes.

When a surge begins, the first indication is thickening of the ice in the upper reaches of the glacier. The subsequent appearance of thousands of crevasses heralds the onset of more vigorous flow. This break-up of the surface spreads rapidly downglacier and the maximum ice movement accelerates from perhaps only a few centimetres a day to as much as 100 metres a day as the zone of surging ice (the surge front) passes by. If the surge front reaches the snout, the glacier advances in dramatic fashion.

In many cases surges result in a glacier advancing several kilometres over a few months, but other surges take some years to be completed. The largest recorded surge was from an ice cap on the Arctic island of Nordaustlandet in Svalbard, when the outlet glacier of Bråsvellbreen advanced 20 kilometres into the sea along a 30 kilometre-wide front some time between 1936 and 1938. Occasionally, however, the surge

Looped moraines in the Roslin Gletscher in East Greenland. A long period of quiescence between surges means that the surface can become relatively smooth and long surface streams can develop. Looped moraines are features of surge-type glaciers, representing irregular pulses of ice from tributaries, and they are the best means of identifying surging glaciers in areas where no records of surging exist.

peters out before the snout is reached, the ice that has been moving forming a bulge just behind the inactive ice of the snout. The 1982/83 surge of Variegated Glacier, for example, ceased a kilometre from its snout, but this is rather unusual.

Surges are accompanied by a tremendous release of meltwater, and the combination of severe flooding and rapid ice advance has been known to cause considerable damage downvalley. Ice is transferred from the high parts of a glacier to the snout, with a fall in the ice surface of as much as 50 to 100 metres in the upper part (often leaving blocks of ice stranded on the hillsides), and a similar rise in the lower part.

When not moving fast, many surge-type glaciers can be identified from the pattern of medial moraines: instead of lying in relatively straight lines more or less parallel to the valley sides, debris is contorted into loops. A loop develops when a tributary glacier flows into the main one while the latter is inactive, and a surge of the main glacier will carry it downwards several kilometres. Repetition of this process in several tributaries can create a complex set of contorted moraines.

Why do glaciers surge?

Several mechanisms have been invoked to explain glacier surges, but the phenomenon is still poorly understood. At first it was thought that earthquake-generated avalanches might trigger them, but more recently scientists have tried to link surges to glacier size, shape, orientation and gradient, climate, bedrock type and thermal regime. However, none of these hypotheses has proved to furnish an adequate explanation.

In recent years two further theories have evolved from detailed observations of surge-type glaciers. Both relate to what happens on the bed, one looking at the distribution of water, the other to the changing properties of deformable sediment. A picture has been built up following studies on the temperate Variegated Glacier, according to which the upper part of the glacier (the 'reservoir area') gradually thickens over a period of time before the surge, while the lower part thins. As a result the glacier becomes progressively steeper, and the stresses at the base of the glacier in the reservoir area increase. Thus the subglacial channels which carry the meltwater close more easily. Eventually the stresses become so great that channels are squeezed shut altogether and the

Left The Variegated Glacier, adjacent to Hubbard Glacier, is known to surge every 16 to 21 years. The 1982–3 surge was therefore predicted and glaciologists were well prepared to undertake the most detailed investigation of a surging glacier ever. Surveying markers were placed on the glacier in order to monitor the velocity, but placing the markers was very difficult because the ice broke up to such an extent that it became impossible to walk on it. Helicopters were used to set markers up, many of which were lost in crevasses and had to be replaced.

Above Three years after the surge of the Variegated Glacier, the glacier is relatively inactive, and the large crevasses and chasms that formed during the surge are rapidly ablating away. However, this late July 1986 picture shows large areas of very rough ice, with snow drifts in the hollows, making walking slow and hazardous.

water is forced to flow out of the channels and spread out as a film across the entire glacier bed. This drastically lowers the friction by separating the ice from much of the bedrock, and the glacier begins to slide much faster. Rapid sliding usually starts in the reservoir area, but as the fast-moving ice impinges against the slower-moving ice downvalley, very high stresses occur and the surge begins to spread. The resulting 'surge front' propagates downglacier as a wave, which if it reaches the snout initiates a sudden and pronounced advance.

After the surge the glacier is less steep. Therefore the stresses decline and once again allow the opening of subglacial channels, at which point the glacier ceases to slide and drops back on its bed. The newly formed channels enable the subglacial reservoir of water to empty, providing the final dramatic flood. At this point in the cycle the ice velocity of Variegated Glacier dropped from 30 to barely 3 metres a day within only 24 hours, but its final flood transported huge amounts of sediment, burying numerous shrubby trees in front of the glacier.

An alternative, and perhaps complementary, explanation has arisen from studies of the Trapridge Glacier in Yukon Territory. This is a thermally more complex glacier than Variegated, since it is cold in part. It is believed that the glacier rests on a bed of soft sediment, mainly till, rather than bedrock. Drainage is by way of channels incised into the sediment during the quiescent phase. As ice builds up in the reservoir

Glaciers with more than two recorded surges

Glacier	Country	Year of surge (number of years between surges in brackets)								
Bruarjökull	Iceland	1625	(95)	1720	(90)	1810	(80)	1890	(73)	1963
Carroll	Alaska	1919	(24)	1943	(23)	1966				
Kolka	USSR	1834	(68)	1902	(67)	1969				
Medvezhiy	USSR	1937	(14)	1951	(12)	1963	(10)	1973		
Nevado Plomo	Argentina	1788	*	1934	(51)	1985				
Variegated	Alaska	1906	**	1947	(17)	1964	(19)	1983		
Vernagt***	Austria	1600	(78)	1678	(95)	1773	(72)	1845		

* 146 years: possibly there were two more surges during this time. The surge of 1788 was not directly observed but can be inferred from a lake outburst.

** There was probably a surge around 1926.

*** The surges prior to 1845 are inferred from lake outbursts.

area, stresses at the base increase and the channels are no longer able to remain open. Water is then forced to flow through the sediment, which is consequently weakened and therefore prone to rapid deformation. A cushion of deforming sediment is formed in which rapid acceleration of the ice becomes possible.

It is possible that surges are caused by both mechanisms, and possibly others. There is still a lot to be learned about why surges take place, but understanding them is crucial in order to be able to differentiate between major climatically-induced advances of the major ice sheets and surge-related advances.

We must especially learn to predict surges where people or installations are endangered in order to be able to take evasive action. In Alaska the Black Rapids Glacier is being monitored because a surge might threaten the Alaska Pipeline and the Richardson Highway. There is good cause for concern, as the glacier has already clearly demonstrated its capabilities in the winter of 1936–37 when, in early December, the occupants of a lodge on the Richardson Highway were startled to see the 3 kilometre-wide front of the glacier bearing down on them. The normally smooth gentle snout had been transformed into a heavily crevassed ice cliff 100 metres high, and the ice was advancing at rates of up to 66 metres a day. If the surge had continued, it would have dammed a major river, severed the highway and demolished the lodge. Fortunately for the occupants of the lodge and the people of Fairbanks, who relied on the road for links with the outside world, the glacier stopped a short distance from the road.

Elsewhere, surges of the Plomo Glacier out of a side valley in the Argentinian Andes led to the formation of ice-dammed lakes in the main valley. The lakes then burst through the ice dam, and created considerable destruction downvalley. During the surge of 1934, seven out of ten bridges and 13 kilometres of the Mendoza–Santiago railway were washed away, causing several deaths. The peak discharge of water in this surge was 3000 cubic metres a second, comparable to the Rhine's discharge at Rotterdam, and fifteen times the normal run-off. Loss of life and extensive damage from similar events has also been reported on occasions from the Karakorum Mountains on the border between China and Kashmir.

6
Nature's Conveyor Belt

One of the most immediately obvious features of mountain glaciers is the amount of rubble and the numbers of huge blocks of rock that litter their surfaces, especially in their lower reaches.

A glacier can be considered as a sort of conveyor belt for rock debris, transporting material from all points along its length towards the snout. This debris is carried most typically on the surface (supraglacial debris) where it forms bands parallel to the sides of the valley, or near its bed in basal ice, but smaller amounts can be incorporated internally from both the surface and the bed (englacial debris), giving the glacier a very dirty appearance.

Surface debris

For a glacier to be loaded with supraglacial debris, it normally has to have exposed rocks along its flanks to provide the source. Thus ice sheets and ice caps carry almost no supraglacial debris, whereas the low reaches of mountain valley glaciers may be almost entirely covered by it. The main source of this is rockfall caused by frost shattering, and by the unstable nature of the hillsides oversteepened by the glacier. Most falls are small, though occasionally entire landslides cover broad expanses of ice, especially in regions prone to earthquakes, such as Alaska, the Andes, the Himalayas or New Zealand. The size of the actual rocks carried by the glacier depends on the rock type: resistant, unbedded rocks like granite create huge blocks, often the size of small houses, but softer ones like shale or limestone invariably break down into small boulders. There are relatively small amounts of sand and finer material mixed in.

Most debris falling down mountainsides becomes caught up with the ice as it slides past the valley sides. Both the line of debris on the surface at the edge of the glacier, and the ridge of debris left behind as the glacier recedes, are known as a lateral moraine. Where two streams of ice join, the two lateral moraines combine to form a single medial moraine, which appears as a line of debris extending towards the snout, now down the middle of the glacier. Commonly, moraines comprise a variety of rock types which reflect the different source areas. When ablation near the snout brings medial moraines of different composition into close proximity, a glacier may take on a multicoloured striped appearance, as

A trail of rusty-weathering angular rock fragments forms a medial moraine on the surface of Gornergletscher, in Valais, Switzerland. The glacier is flowing downhill away from the camera, towards the Matterhorn in the background. The angular nature of the stones is characteristic of debris transported on a glacier's surface.

Left Large, leaping and rolling boulders leave a cloud of dust in a rockfall on the slopes of Mount Cook above the Tasman Glacier in New Zealand's Southern Alps. Steep rocky mountainsides characterize valleys that are, or have been, occupied by glaciers. They are particularly prone to rockfall, so providing the material which is transported on the glacier surface.

A fine set of narrow medial moraines on a glacier on the northern slopes of the Karakorum Range in Xinjiang Province, China. Ice flow is towards the top left of the picture.

in the justly named Variegated Glacier in Alaska or the Unteraargletscher in the Alps.

Dark debris on the surface of a glacier absorbs the sun's radiation better than the surrounding ice. The debris cover becomes relatively warm, and increases the melting of the ice, particularly if it is thin, with the result that the moraines occupy depressions. In contrast, a thick continuous cover of debris normally slows down ablation, so the debris stands proud above the general glacier surface as a distinct ridge.

Isolated boulders also protect the ice from melting, so they may end up sitting on perches of ice. These 'glacier tables' often tilt towards the sun as time goes by, until they slide off, and the process is repeated.

Streams on the surface may recirculate much of the finer debris, some of which collects in hollows. Once a stream course has been abandoned, the surface continues to sink. However, the debris in the depression may so slow down the melting of the ice below it that sand or gravel comes to sit on small rises, and as the surrounding ice carries on melting these can become dirt cones. If you hack at them with an ice axe you find that they consist simply of a veneer of debris a few centimetres thick covering a cone of ice.

Right Large boulders on the ice surface frequently protect it from the effects of solar radiation, as a result of which they often end up resting on pedestals of ice. This 'glacier table' has developed on Griesgletscher in the Swiss Alps.

Far right Different coloured lines of debris denote different source areas, even though they have merged together to create a continuous cover of debris in the lower part of this, the Unteraargletscher in the Bernese Oberland of Switzerland. The prominent line above the glacier, separating bare rock from ground richer in vegetation, is a trimline, marking the position of the Little Ice Age glacial maximum in the eighteenth century. The peak in the background is the Lauteraarhorn (4042 m).

The proportion of debris cover increases towards the snout in most valley glaciers. Uneven melting of ice beneath the debris cover and the action of streams on slow-moving or stagnant ice create an irregular surface of sharply defined hills and valleys with a relief of several metres. The debris is often unstable and, although not technically difficult to walk on, nevertheless can be hazardous. During the ablation season, debris is continuously falling down the ice slopes, many of which have only a thin cover of debris. Such terrain, with its deeply incised channels, pools and englacial streams, is referred to as glacier karst by analogy with similarly scarred areas of limestone.

If, as a result of climatic warming, the glacier wastes away, supraglacial debris is lowered on to the bed beneath, or on to debris being deposited from the base of the glacier. Such deposits are known as supraglacial meltout tills (till being the term for the poorly sorted deposits released directly from ice). The deposits are irregularly spread over the ground and are constantly reworked by sliding and slumping, and by streams, until sufficient vegetation grows to stabilize them.

Debris transported along the glacier bed

Debris at the bed of a glacier is very different in character from supraglacial debris, since it is continually being modified as the ice moves downhill.

Before a glacier can pick up debris the ground may have to have been deep-frozen and thawed, processes which loosen blocks of rock, especially along bedding planes. The blocks can then easily be incorporated by freezing on to the base of the glacier as it slides on its bed. Further debris is collected as a result of small-scale changes in pressure around bedrock bumps, which melt and refreeze the ice: as it passes over a rise, the pressure rises, thereby lowering the melting point; then on the downstream side the pressure lowers and the water refreezes, producing regelation ice. This process can be seen at home by making a weighted wire cut its way through a block of ice – the cut is sealed over as the ice freezes again after the wire has passed through it. Many glaciers have a layer of debris-rich ice created this way, known as the sole.

Now detached from the bed, the blocks of rock become powerful tools for eroding the bedrock further. Held firmly in the ice, the rock fragments groove and scratch (striate) the bedrock, and themselves have their sharper corners broken off. They also rotate within the ice because the velocity of the glacier increases upwards, so new parts of the blocks are constantly being exposed to the bedrock. Thus a rock that started

Longitudinal profile through a retreating valley glacier, illustrating debris in transport and some of the varieties of glacial sediment that result from deposition.

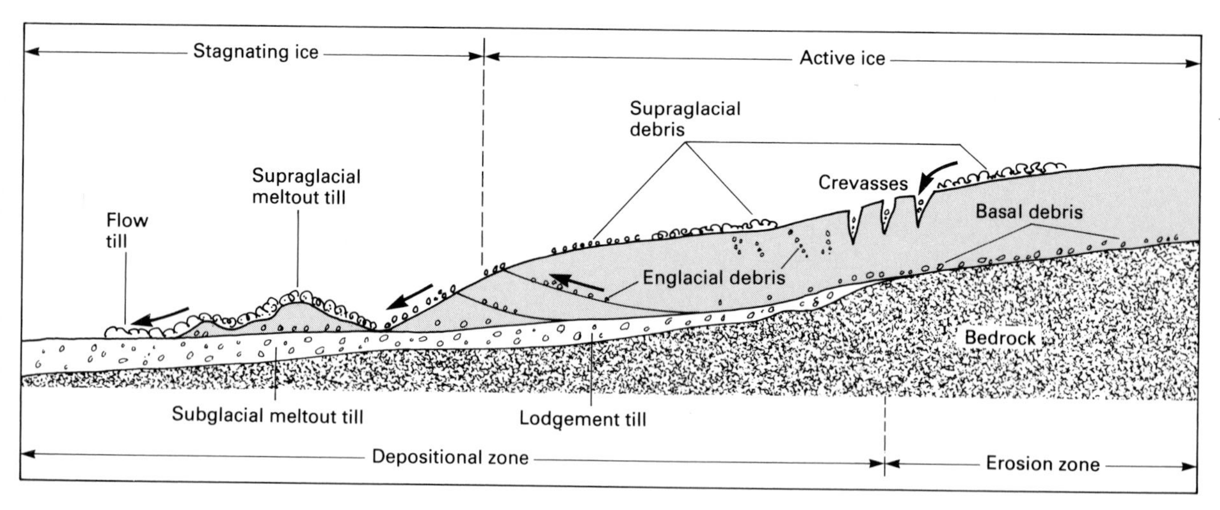

angular becomes progressively more rounded and acquires striae, some stones becoming quite round (although not as round as those carried by streams). Most basal debris is picked up beneath the accumulation area where the ice is more dynamic, or along the valley sides. It is in these areas that the sandpaper effect of the glacier sole is greatest and most erosion occurs. In the ablation area, where the glacier is slowing down, more of the basal debris load is deposited.

These processes of grinding and crushing at the glacier bed also generate a much finer material, including clay and silt – the 'rock flour' which gives the characteristic milky appearance to glacial streams.

The processes of deposition are complex and involve meltwater, but two main types of basal till result from it. At first, with the ice deforming rapidly, melting near its bed and sliding rapidly, debris is actively plastered on to the bed to give a 'lodgement till'. The lodgement process becomes less effective towards the snout and direct melting out of debris from the ice takes place, producing a 'meltout till'. Both types of till contain a random mixture of material from clay to boulder size.

Sheets of till cover much of North America and Europe, where they provide fertile ground for agriculture, since the resulting soils are rich in

Although some cold Antarctic glaciers are frozen to the bed, others are thick enough to overcome the effects of sub-zero atmospheric temperatures and slide on a film of meltwater. In such conditions debris can be eroded and picked up by the glacier. This November 1986 photograph shows the base of the margin of Taylor Glacier in Victoria Land, Antarctica. This base is characterized by evenly distributed debris, above which is a series of parallel debris-rich layers, each being a separate regelation event, characterized by melting and refreezing. Ice flow is from left to right.

Far right Like the boulders that form glacier tables, piles of debris which collect in depressions on the ice surface also protect the ice from ablation, eventually creating 'dirt cones', like this 2-metre-high example on Gulkana Glacier, Alaska. The debris cover may only be a few centimetres thick, as in this case.

minerals. However, tills sometimes generate problems for the construction industry because, although providing a firm base for much of the time, when wet they can deform easily, and on slopes can be subject to landslippage.

Debris inside a glacier

A certain amount of debris becomes incorporated in the interior of a glacier as englacial debris. Surface debris from rockfalls in the accumulation area becomes buried by snow, and other debris falls down crevasses, though in both cases it remains below the surface until released by ablation. Debris may also be found in an englacial postion along thrusts, which are faults extending upwards at a low angle from the bed. Some of this debris may even reach the surface, especially near the snout, but its basal origin can be recognized from the part-rounded and striated nature of the stones.

Right In addition to being transported at the base and on the surface of a glacier, debris is also carried within the body of the ice 'englacially'. Such debris may have fallen from the surface down crevasses, or it may have become incorporated into the ice by folding and thrusting of the debris-rich basal layers. The latter process can be seen here operating in the Taylor Glacier, Victoria Land, Antarctica. In the foreground is a small ridge of debris-rich ice which is releasing its debris by sublimation (evaporation of ice without melting). The frozen Lake Bonney on either side of the ridge is highly saline with a sub-zero melting point; it is characteristic of lakes in this polar desert region.

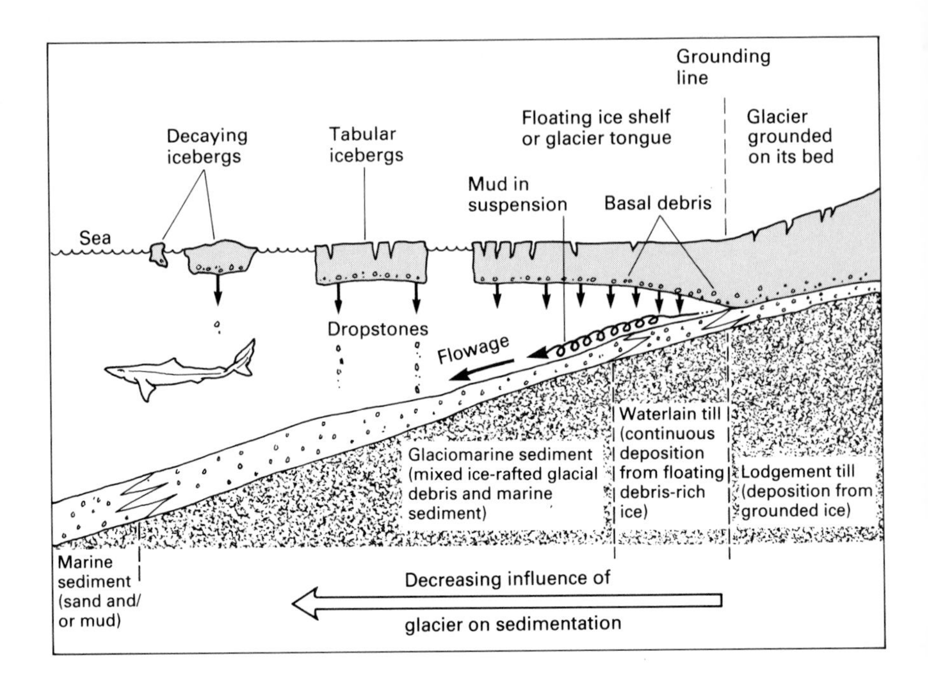

Longitudinal profile through a floating tidewater glacier, illustrating debris transport, depositional processes and the resulting sediments.

There seem to be some differences in the amount of debris carried according to the temperature of the glacier. Cold glaciers tend to have a very thick basal debris load, perhaps representing 50 per cent of the total ice thickness towards the snout. This is probably due to a combination of freezing of debris to the base and folding in the basal part of the glacier, repeating particular debris layers. In contrast, the amount of debris carried on the surface is small. The reverse is true for temperate glaciers. These commonly have an extensive debris cover, though the basal debris layer is normally only a few metres thick. Many scientists have argued that cold glaciers which are frozen to their beds tend to be inactive in terms of their effect on the landscape, but rock fragments embedded in the ice may erode the bed to a limited extent, despite the lack of sliding. Furthermore, there is some evidence that the large ice sheets of North America and Eurasia have ripped up large blocks of rock from the bed even if it were frozen.

Debris in glaciers also includes material that has been carried by the wind, and most glaciers have minor amounts of wind-blown silt derived from glacial rock flour deposited by streams beyond the confines of the glacier. Sometimes the debris originates in storms far distant; for example, every couple of years or so, yellow Saharan dust is blown onto Alpine glaciers, occasionally staining the glaciers noticeably. In other places, notably Iceland and Alaska, volcanic eruptions have thrown significant amounts of ash on to a glacier, and if the time of the eruption is known, the ash layer can tell glaciologists a lot about glacier mass balance and dynamics.

Other types of debris in ice are volumetrically insignificant, but are important in providing clues about environmental change, such as traces of industrial pollution or forest fires, and radioactive fall-out from nuclear explosions. Studies of the acidity of snow and ice have also provided much information about past volcanic eruptions, even when the actual dust and ash have been invisible.

7
Ice and Water

Glacial meltwater plays an important role in the landscapes around glaciers, and in some cases in the lives of the people who live near them. In Switzerland, up to 15 metres of ice melt from the lower lying parts of the Grosser Aletschgletscher every summer and help generate the hydroelectric power used by the totally electrified Swiss railway system. In the arid regions of north-western China and in Argentina glacial meltwater irrigates desert land, ensuring the survival of many thousands of people.

Meltwater also plays a significant part in the development of glacial landscapes, even in high polar regions where, during the short summer season, snow and ice melt combine with rainwater to provide a noisy, tinkling, gurgling, rushing or roaring background to any activity near glaciers. Glaciers everywhere generate their own stream systems, either on their surface or within and below the ice, in a similar manner to streams in limestone regions. During the peak period of melting in early summer the stream that emerges at the snout of a glacier is often a spectacular torrent, frequently flooding the valley floor below. Yet in winter, discharge is reduced to a mere trickle and in many parts of the world water is locked solid for as many as nine months of the year. These extremes between summer and winter provide a fascinating range of meltwater features on and around glaciers.

Factors affecting melting

The principal influence on melting is, of course, air temperature, although even on bright sunny days with sub-zero temperatures, melting may be significant because of intense radiation. Outside Antarctica, practically all glacierized areas experience daytime summer temperatures several degrees above freezing. During fine weather the sun's radiation generates large volumes of meltwater, but at night there is usually an almost total freeze-up. Consequently, discharge fluctuates on both a daily and seasonal basis. Extremes of daily discharge become more pronounced as distance from the poles increases, whereas seasonal extremes are more pronounced in high latitudes. A practical aspect of these daily variations is that, whereas a meltwater stream may be easily crossed in the morning, by mid-afternoon it may be impassable, and it is therefore quite easy to get stranded on the wrong side. In cloudy weather

Summertime at the side of Steingletscher, near the Susten Pass in Switzerland, illustrates the close association of meltwater, ice and rock.

Undercutting of the Wordie Gletscher ice cliff in northern East Greenland causes frequent collapses in the cliff, temporarily blocking the stream and creating lakes. The discharge of the main river below the glacier thus fluctuates widely, and has numerous fragments of ice floating in it.

daily extremes are not so pronounced and, although daytime meltwater production may be reduced, melting will continue at a slightly lower level during the night.

Another agent promoting melting is geothermal heat – the heat from the earth's interior. In its most extreme form, such as in volcanic regions, geothermal heat can melt large volumes of ice and create sub-glacial lakes, but normally it only produces limited melting at the glacier bed. Nevertheless, such basal melting is a significant process beneath many otherwise cold glaciers. Although the bulk of the ice may be at sub-zero temperatures, and in the absence of geothermal heat would reflect the mean annual air temperature at the altitude of the site measured, geothermal heat causes the temperature to increase with depth in the glacier; if the ice is thick enough, melting occurs at the base. In such a case, the glacier behaves like an insulating blanket over the land: quite moderate temperatures are found under the glacier, while the ground around the glacier may be permanently frozen to a depth of several hundred metres.

Snow swamps

The pattern of melting differs between cold and temperate glaciers. The former often do not have a well-defined firn line in summer, and it is common for large areas of the winter snow pack to become saturated with water in early summer because the cold ice prevents a good internal drainage network developing. These areas of saturated snow are known as snow swamps. They are treacherous because their surface may appear like any area of dry snow, but beneath may lie metres of slush or even streams. They are often unstable, and it is common for any disturbance, such as a person walking across them, to trigger a slush avalanche, or slush flow. Usually such flows are small, but occasionally large areas of the snow pack become active and whole areas of a glacier are swept clear, exposing bare glacier ice. As the saturated snow comes to rest and the water escapes, the slush packs into a very hard mass of dense snow. If a saturated snow pack remains in place, the first indications of meltwater will be vague channels which appear deceptively small. Anyone unfortunate enough to fall in might be extremely difficult to get out.

Meltstreams and slush on the surface of a glacier may be hidden by a superficial cover of snow that can form a trap for the unwary traveller, as here on the White Glacier, Axel Heiberg Island in the Canadian Arctic.

Far right Some moulins, created by large streams, reach depths of 100 metres and widths of 10 metres or more. Like this one on the Mer de Glace in the French Alps, they can be even more intimidating to the walker than large crevasses.

On temperate glaciers the firn line is well defined and normally only a narrow zone of saturated snow develops. Stream courses are also well-defined from an early stage in the melt season, because it is easier for meltwater to drain downwards to the bed than it is on cold glaciers.

The glacier drainage system

The development of meltwater channels on the surface of a glacier depends on the rate of melting, the rate of deformation of the ice, the extent of crevasses and the pattern of other structures such as foliation, and ice temperature. Surface channel systems develop best on stagnant and on cold glaciers, but will not appear at all on those with a large number of crevasses. The channels themselves range in size from tiny rills to canyons several metres deep and a hundred or so metres wide that form an impassable obstacle when a stream is flowing. On flat, crevasse-free glaciers, the streams may form into a dendritic pattern – like branches of a tree joining to form the main stem. Alternatively, they may form tight and meandering patterns, with deeply incised channels marked by undercut walls on the outside of bends.

In slow-moving or stagnant ice, streams have time to evolve over many summers, and they may develop complex meandering courses. These irregular forms were photographed from a helicopter above Vibeke Gletscher in northern East Greenland. The field of view from left to right in the bottom of the picture is approximately 300 metres.

Glacier drainage is commonly influenced by the distribution of surface debris and structures such as foliation. Medial moraines often stand out as ridges within a larger longitudinal depression. Streams may also flow parallel to foliation, since different ice types melt at differing rates, forming a characteristic ridge-and-furrow topography. Glacier structure controls drainage in a still more fundamental way. Ice structures, especially the traces of former or developing crevasses, act as planes of weakness which are exploited by meltwater to form a glacier mill or *moulin*, similar to a pothole in limestone country. These range in diameter from a metre or less to as much as ten metres. It is through them that much of a glacier's surface meltwater reaches the bed, or at least an internal drainage network, and a view from the edge of a moulin

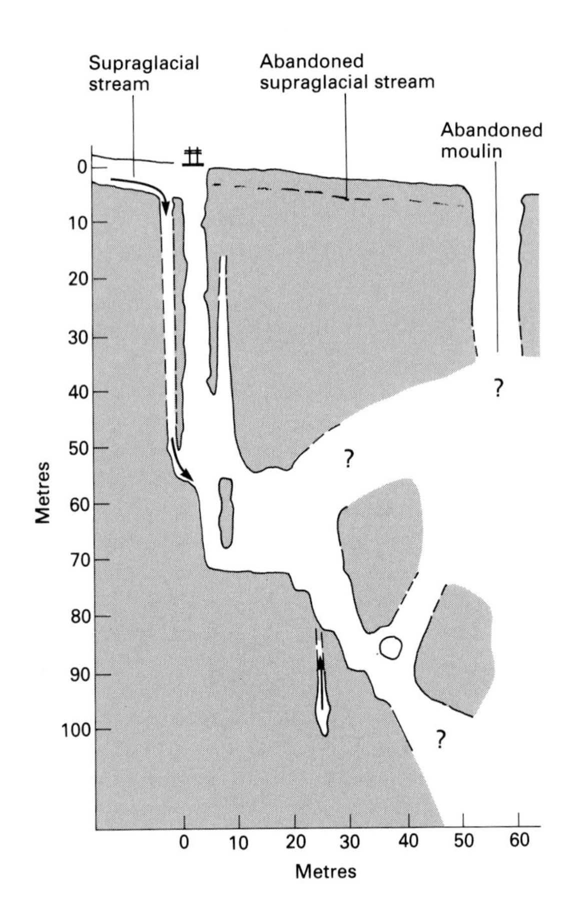

Vertical cross-section through a large moulin in the Mer de Glace that was explored in 1986.

Water in the interior of a glacier is sometimes subject to high pressure, and on rare occasions may burst out at the surface in fountains. This small one on the Oberaargletscher, Switzerland, lasted just a few seconds before subsiding.

Right Partially drained ice-dammed lake with numerous stranded icebergs in a valley adjacent to the Vibeke Gletscher in northern East Greenland.

Below This lake on the Gornergletscher near Zermatt, Switzerland, is one of several that have developed in the cold ice that descends from an exceptionally high elevation near the summit of Monte Rosa.

down which a large meltwater stream falls presents an awesome sight.

Glacier surface streams are often difficult to cross. Although narrow, the ice-cold water and the slippery sides and bottom of a channel make jumping hazardous, and one may have to walk many kilometres to get round them. Indeed, unlike the case with normal streams it is often better to walk *downstream* to cross one as there is a good chance that the water will disappear down a moulin.

Small pools of standing water may develop on the flatter parts of a glacier, especially where dark patches of debris or dust absorb more solar

radiation and melt down into the ice. The smallest examples, known as cryoconite holes, are cylindrical tubes a few centimetres across, but maybe tens of centimetres deep. One gets the impression that the debris in the bottom has bored its way down into the ice. In sunny weather cryoconite holes develop in large numbers and occasionally give the surface of a glacier a honeycombed appearance. The holes merge to form larger ponds, and in extreme cases, lakes tens of metres across, for example on Gornergletscher in Switzerland. Such features normally have near vertical sides, and their levels may fluctuate between day and night if they are linked to the internal plumbing system of the glacier. Large lakes seem to develop exclusively on cold glaciers.

Subglacial streams often emerge from the snout of a glacier via an open cave, known as a 'glacier portal'. The snout of the Gornergletscher, Switzerland, shows the typical sediment-rich nature of such streams.

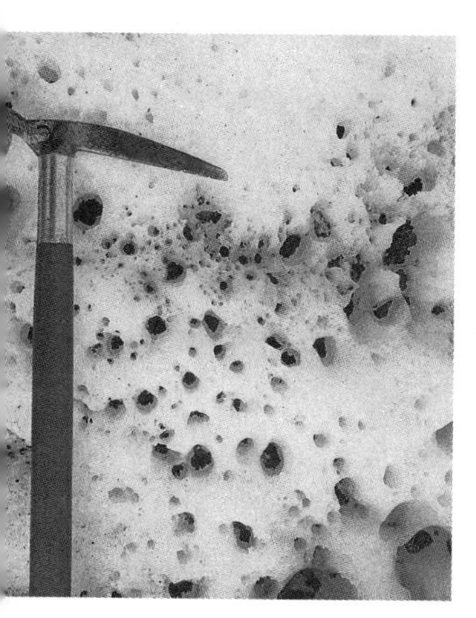

'Cryoconite holes' on the surface of Griesgletscher, Switzerland. They are formed by minor accumulations of dust and other debris absorbing more radiation than the ice surrounding them, and then melting down a few centimetres into the ice surface.

Not a great deal is known about the internal drainage systems of glaciers. Nevertheless, through the use of dye tracers it has been possible to monitor how fast water goes through them. For example, on the north Norwegian glacier Austre Okstindbreen typical rates of water throughput were recorded as 0.6 to 0.8 metres per second (22–29 km per hour) over distances of 500 to 1000 metres. However, peak rates of 1.8 metres per second (65 km per hour) have also been documented. Studies on the Haute Glacier d'Arolla in the Swiss Alps have shown that the internal drainage system there opens up as the melt season progresses, as recorded by the rate at which dyed streams pass through the system.

On another Norwegian glacier, Charles Rabots Bre, a sudden increase in discharge at the snout was noted within only twenty minutes of the peak of a thunderstorm, indicating a very open internal drainage network. In other glaciers meltwater may be held back in internal reservoirs and released more gradually. The difference reflects the relative sizes of englacial and subglacial channels.

Temperate glaciers have very different internal drainage systems from cold ones. In them most water reaches the bed well before the snout, and emerges from a single glacier portal. In contrast, most meltwater migrates towards the margins of cold glaciers because it cannot easily penetrate sub-zero temperature ice. There, it cuts deep channels between the hillside and the ice. The snout of a cold glacier rarely has a glacier portal, instead the marginal meltwater streams may flow around the snout edge, sometimes to merge before flowing off down the valley. Once again, therefore, meltwater streams prove to be more of an obstacle to crossing a glacier in polar regions than in temperate ones.

8

The Birth of Icebergs

One of the finest sights in nature is a large, heavily crevassed valley glacier descending from high mountains into the sea, where glistening blue icebergs break away from the vertical, calving cliff face. Icebergs then disintegrate in irregular fashion. They tilt (as seen from tidemarks running diagonally up their sides) and often turn turtle. As they melt away, they develop fascinating and beautiful profiles, sometimes resembling castles, churches, arches and other architectural shapes. An iceberg-filled fjord will resound to the sharp thunder-like noise of an iceberg cracking, the swish as it turns over, and the crackling and popping as pressurized air bubbles in the ice escape.

Such glaciers which terminate in the sea, called tidewater glaciers, are only found in high latitudes: Greenland, the Canadian Arctic, Svalbard, the Soviet Arctic, Alaska, Chilean Patagonia, South Georgia and Antarctica. Apart from those who travel on cruise ships to such areas, few ordinary people gain a first-hand acquaintance of the beauty of tidewater glaciers.

Grounded tidewater glaciers

Of the two main categories of tidewater glacier, grounded tidewater ones are typical of the fjords of southern Alaska and Svalbard, the Canadian Arctic and Greenland. As the name suggests they rest on a bed, which may be a hundred metres below sea level or more, while an often heavily crevassed terminus forms a vertical cliff which may rise 50 or more metres above the water. The forward motion of the glacier creates a very unstable face, with the result that large pinnacles and enormous blocks of ice frequently topple and crash into the water, creating magnificent displays as fountains of water shoot high into the air. Sometimes calving occurs from the ice mass below the water, giving rise to the unnerving sight of blocks of ice being projected up out of the water. It is accompanied by unpredictable waves, which can come as a shock to anyone camping close to the shore, or sailing too close to the glacier.

Most tidewater valley glaciers are situated well inside fjords, having retreated rapidly since the Little Ice Age of the eighteenth and nineteenth centuries. Although they do produce icebergs, these are

Previous page Floating tidewater glaciers commonly produce tabular icebergs, which decay as they drift away. The top of this castle-like berg in Antarctic Sund, East Greenland, is rough because of original crevassing, and it has an inclined tidemark that illustrates how the buoyancy of the berg is adjusting to melting.

Left The Cascade and Barry glaciers in Harriman Fiord, Alaska, descend from the high mountains and converge at the coast. Many such tidewater glaciers in Alaska produce large numbers of small icebergs.

relatively small (up to a few hundred metres or so across) and only a few reach the open sea. Many of them become stranded on banks, especially in areas of high tidal range where they may leave grooves as they are dragged across the sediment.

As glacier ice melts in water, the air bubbles trapped under pressure for hundreds of years are released, making a crackling, popping noise, reminiscent of a well-known breakfast cereal. In North America this noise is referred to as ice sizzle.

In parts of the high Arctic, especially in Nordaustlandet in Svalbard, grounded tidewater glaciers form ice cliffs or ice walls stretching for many kilometres. In much of the Antarctic grounded ice cliffs extend much further, for hundreds of kilometres, but, being slow-moving, they

Right Torrellbreen is a large glacier that descends to the west coast of Spitsbergen from highland icefields at an elevation of 1000 metres. The glacier is grounded on the sea bed and produces only small icebergs. The nunatak of Raudfjellet (1014 m) is in the background; in the foreground wave action is sorting the muddy boulder-strewn glacial sediment, forming small sand bars.

Above The Columbia Glacier in Alaska – a good example of a tidewater glacier that relied on a bedrock sill as an anchor, but, having detached itself, is now retreating rapidly.

Hubbard Glacier in southern Alaska has always been an actively calving glacier, even before the rapid advance of 1986 (see chapter 5), as in this 1983 photograph. The glacier tongue rests on the sea bed, and the constant action of waves undercutting the ice leads to frequent collapses of the cliff.

account for only a small proportion of the ice discharge into the Southern Ocean. Ice cliffs are in any event normally less active than grounded valley glaciers, have fewer crevasses and a cleaner, more regular cliff line; they also produce icebergs at a slower rate.

Some tidewater fjord glaciers are grounded only on a shallow sill of bedrock or moraine, behind which the rest of the glacier floats above much deeper bedrock. If such a glacier retreats off its sill it is likely to continue retreating extremely rapidly. A good example is the Columbia Glacier in Prince William Sound, Alaska, which behaved just as glaciologists had predicted when it had retreated and thinned sufficiently for the pinning effect of the sill at its terminus to be removed. By the late 1980s, the glacier front had backed off into deeper water behind the

Above A marine ice tongue surrounded by winter sea ice on the coast of northern Victoria Land, Antarctica. Such tongues are the floating seaward extensions of valley glaciers and ice streams. Where they cease to be constrained by valley walls or less active glacier ice, they develop fin-like features. As a glacier tongue extends seaward, the effect of tides causes it to bend, and eventually to break off. The resulting icebergs occasionally exceed 100 kilometres in length.

Above right The birth of a small tabular iceberg from a floating ice embayment in Coats Land, East Antarctica.

Right Cold Antarctic tidewater glaciers move much more sluggishly and are less crevassed than the temperate glaciers of Alaska. The Ferrar Glacier in Victoria Land, Antarctica, illustrates how large tabular icebergs are born from them. In the foreground is old blue sea ice. The giant groove, probably 50 metres across, running up the glacier is a large meltwater channel, which facilitates break-up.

sill and its instability led to retreat at a rate of several hundred metres a year, accompanied by a vastly increased supply of icebergs. Sadly, the simultaneous prediction that the resulting icebergs might create a hazard to shipping was fulfilled catastrophically in March 1989 when the *Exxon Valdez*, one of the oil tankers that carry oil away from the terminal of the Transalaska Pipeline at Valdez, ran aground, according to some reports trying to avoid ice from the Columbia Glacier. More than ten million gallons of oil poured into Prince William Sound from the holed tanker, resulting in the USA's worst ever oil pollution incident. The marine life of the western part of the sound was decimated as the oil drifted ashore.

Floating tidewater glaciers

Other glaciers which enter the sea float, because they flow into deeper water, especially in Greenland, the Canadian Arctic and Antarctica. In contrast to land-based glaciers, which slow down towards the snout, floating tidewater glaciers accelerate as soon as they begin to float, and as a result the tongue is normally heavily crevassed. Icebergs from them consist not only of pieces of ice that have fallen from the calving cliff, but of entire sections of the tongue, known as tabular icebergs. Those in the Arctic, where they usually originate in the fjords of Greenland, frequently measure several hundreds of metres across.

The Antarctic has numerous floating 'ice tongues' tens of kilometres long which project from the coastline, unconstrained by valley walls. They are the extensions of ice streams or valley glaciers on land which develop fin-like margins, and they grow until the effect of the bending under the influence of tides makes them unstable. Then they break up and form huge icebergs.

Ice shelves

Ice shelves are found where glaciers are generated at sea level or descend from the mountains, and discharge into the sea, coallescing on a broad front. In the Arctic they only occur in Ellesmere Island, but nearly half the Antarctic coastline is made up of them. They are self-generating

Right A tabular iceberg derived from the Ekstroem Ice Shelf in western Dronning Maud Land, Antarctica, drifts slowly by the German research vessel FS *Polarstern*.

at sea level there, much of the ice forming *in situ* from snowfall, with ice flowing down from the mountains making up a smaller proportion. Ice shelves produce the largest icebergs, and some detached from Antarctic ice shelves are many kilometres across. One of the biggest recorded broke away from the 800-kilometre-wide Ross Ice Shelf in 1987 and was nearly 160 kilometres long, with an area of over 6250 square kilometres. It was said by a spokesman of the US National Science Foundation to contain enough water to supply Los Angeles for 675 years!

Iceberg drift

Icebergs do not necessarily float in the direction of the surface current, but may be influenced by deeper currents or by the wind. Thus a large

Stranded icebergs, in various stages of decay, resting on the 400-metre-deep sea bed, and surrounded by sea ice, in the eastern Weddell Sea, Antarctica.

A typical Antarctic tabular iceberg, with prominent blue ice layers defining the annual layering, in the Lazarev Sea, East Antarctica.

The final stages of decay of an iceberg are represented by small icebergs called 'growlers' such as this one in the Lazarev Sea, Antarctica. As they are continually awash with water, they are often difficult to spot, and are therefore a serious hazard to smaller vessels.

iceberg may travel in the opposite direction to smaller ones. And some seem to have minds of their own. On a scientific drilling project in the southern Indian Ocean, one of us came across one that was towed away from the ship, as it posed a threat to it, but it floated back. It was then towed in a different direction, but again floated back. It was then towed away in yet another direction, but once again it floated back! Meanwhile, other nearby icebergs kept a steady drift in a single direction.

If caught up in the appropriate currents, icebergs can travel for thousands of kilometres before breaking up. Many icebergs from the west coast of Greenland circulate around Baffin Bay, and some travel as far as Newfoundland, so creating a hazard to shipping in the North Atlantic and to drilling platforms on the Newfoundland continental shelf. The most famous event associated with an iceberg collision was the tragic loss of the 40,000 tonne *Titanic* south-east of Newfoundland in April 1912, when 1503 people were drowned. The iceberg was described as standing 24 metres above the water, with a length of 18 metres, and was estimated to weigh 200,000 tonnes.

Icebergs from Antarctica tend to circumnavigate the continent in a westerly direction before becoming caught in easterly currents further out from the coast. At times, notably in the last century, icebergs have been sighted as far north as the shipping lanes off the coast of southern Africa, probably originating in major calving events from ice shelves.

9

Ice and Fire

In 1986 the volcano Nevado del Ruiz in the Colombian Andes erupted. On a geological scale, the event was rather small and no major explosive blast occurred. Only relatively small amounts of ash were deposited, mainly on the eastern side of the volcano, and little physical damage was caused. Nevertheless, around 30,000 Columbians lost their lives. Nevado del Ruiz had triggered the most tragic of all twentieth century volcanic catastrophes, and it would not have happened but for the volcano's icecap and snow cover. When the volcano began to erupt, glacier ice melted and the resulting waters carried away enormous quantities of recent and older ash, creating a mud flow, or lahar. The lahar tore down the mountainside into the densely populated valley below, burying the town of Armero completely. For several days tragic scenes from Armero appeared on television screens around the world, showing rescuers attempting to free people buried in the dense mud.

In many parts of the world, volcanoes, snow and glaciers combine to form some of the most spectacular, but potentially also very dangerous, landforms. Snow is common on Italy's Mount Etna (3340 m), where ski lifts provide Sicily's only winter sports facilities, on Japan's Fudji-san (3776 m) and even on Hawaii's Mauna Loa (4171 m) and Mauna Kea (4206 m). Glaciers drape dozens of volcanoes in the Andes of Columbia, Ecuador, Peru, Bolivia, Chile and Argentina. The highest mountain in Africa, Kilimanjaro (5895 m), still bears a few small summit glaciers, as do certain other mountains in East Africa. The world's southernmost active volcano, Mount Erebus (3794 m) in Antarctica, is almost entirely glacier-clad, yet contains a lava lake in its summit crater, thereby providing one of the greatest natural temperature contrasts within a few metres in the world (around 1000°C for the lava and −60°C or lower for the winter air temperature).

The North Island of New Zealand has several impressive volcanoes, the highest of which, Ruapehu (2797 m), provides the country's premier skiing area. Ruapehu receives a heavy winter snowfall, and carries a small glacier: on Christmas Eve in 1953, an outburst from a crater lake, supplemented by melting snow and ice, created a lahar which swept away a railway bridge shortly before the packed Wellington–Auckland express was to pass by. The locomotive and five carriages plunged into the torrent and 151 people perished.

Previous page A late evening view of Mount Rainier (4392 m) in Washington State. The volcano has erupted frequently in the past 1000 years, but not this century.

America's Cascade range is crowned by a whole string of glacierized volcanoes. On Mount Shasta in the Californian part of this majestic mountain chain we find the southernmost glaciers on United States' territory. Even Mexico's Popocatapetl (5452 m) and Citlaltepetl (5700 m), the country's highest mountain, are crowned by glaciers. In the USSR glaciers are found on volcanoes on Siberia's Kamchatka Peninsula and its geological continuation into the Aleutian Islands, and from there to mainland Alaska.

More glacierized volcanoes exist in the Southern Ocean and the North Atlantic. Iceland probably has best documented history of floods caused by volcanic activity beneath glacier ice. Ten per cent of the country is covered by ice caps, most of which lie astride a volcanically active rift zone along which the earth's crust spreads at a rate of several centimetres a year.

Below The ice-draped Mount Erebus (3794 m) seen rising above the Bowers Piedmont Glacier in McMurdo Sound, Antarctica, is the world's southernmost active volcano. Winter sea ice extends between the glacier and the volcano, a distance of some 50 kilometres.

Breidamerkerjökull is the outlet glacier of the ice cap of Vatnajökull in southern Iceland. It terminates in the coastal lagoon of Jökulssarlon, where it produces small ash-laden icebergs. In the background the irregular ash layers in the glacier itself testify to numerous volcanic eruptions in the past.

The largest of Iceland's ice caps is Vatnajökull, measuring 150 kilometres in length and averaging 420 metres deep, though reaching a maximum depth of 1000 metres. Near its centre is a volcano, the Grimsvötn Caldera, which has erupted frequently in historical times. Sometimes ash eruptions have been observed by airline pilots on routine flights, and even small eruptions are prominent if they penetrate the ice since the ash contrasts well with the firn. The lake in the caldera reaches a depth of 300 metres and when this bursts out from the ice it creates huge floods called *jökulhlaups* (an Icelandic term meaning glacier runs) to the south of the ice cap; in 1934 approximately seven cubic kilometres of water flowed out this way. Prior to 1934 Grimsvötn burst approximately every ten years, but since 1954 the interval has shortened to an average of five years, with events occurring in 1954, 1960, 1965, 1972, 1976 and 1982 – not always because of eruptions, but often just because of the high heat flow close to the volcano. These floods have created the huge, flat, braided river plains, which until recently defeated all attempts to construct a safe permanent road across the area.

Another subglacial volcano is Katla, which underlies Iceland's

Two volcanoes in Chile's Parque Nacional Lauca have contrasting appearances: Volcan Parinacota (6330 m; right) has been more recently active than its northern neighbour, Volcan Pomerape (6240 m), and glacial as well as stream erosion has progressed less far. Accordingly Parinacota has a more regular cone than Pomerape.

The eruption of glacier-bearing Mount St Helens in 1980 had a devastating effect on the neighbouring landscape. This view shows the breach in the northern rim of the mountain. The gully below still contains remnants of the former Shoestring Glacier, largely covered by debris. During the eruption, the glacier's accumulation area was blasted into the air. Glacial meltwaters mixed with volcanic debris to create a lahar that rushed down the northern side of the volcano, destroying the forest in the foreground. Eight years after the eruption, many flowers could be seen re-colonizing the scene of destruction.

southernmost ice cap, Myrdalsjökull. Jökulhlaups from Katla have been recorded since 1625, the last event taking place in 1918 when an immense flood poured from the glacier during a violent eruption. The flow rate is estimated to have reached 200,000 cubic metres per second, about the same as the discharge from the Amazon.

The eruption of Mount St Helens

Perhaps nowhere have the effects of a volcanic eruption been studied more intensively than on Mount St Helens in the Cascades of Washington State. Following a huge eruption on 17 May 1980 the mountain's height reduced from 2949 to 2549 metres and most of its glaciers disappeared. The earliest activity had been on 27 March 1980, when after several days of earthquake activity, small ash and steam explosions penetrated the summit ice cap, ending the volcano's 123-year dormancy. During April and early May magma appeared on the previously symmetrical cone, producing a noticeable bulge on its northern side. However, few ice avalanches were generated at the time, despite the oversteepening of the mountainside, and the glaciers remained more or less intact.

The huge subsequent eruption on 17 May resulted from slope failure on the bulging north side of the mountain, which created a huge landslide. Sequential photographs of the event show an enormous ice avalanche tearing down the mountain, probably representing most of the accumulation area of Forsyth Glacier, which had been thrown off the summit area as the explosion commenced. During the eruption 70 per cent of the mountain's glacier volume disappeared. The glaciers on the eastern, western and southern sides of the mountain partially survived the eruption, but because the volcano blew off its summit area, they lost most of their accumulation areas. Yet, during the next few years the glaciers advanced as they had been covered by a blanket of ash which protected them from solar radiation. However, these glacier remnants are now shrinking and soon little or no ice will be left on what was previously one of the most graceful volcanoes in North America.

It was clear to geologists that any major eruption would be accompanied by large lahars. As it happened, an evaluation of the risks of mudflows had been conducted a short time before the mountain resumed activity in 1980, and precautions were taken to minimize damage by lahars, including the lowering of one of the hydro-electric reservoirs on the Swift River. During the eruption, it was possible to contain in the reservoir not only the floods from the south flank of the mountain, but also many thousands of trees that had been felled by the lahars. In other rivers, on the other hand, trees jammed under bridges dammed the flood until the bridges were pushed from their foundations and destroyed.

Whereas only a few square kilometres of ice had covered Mount St Helens before the eruption, its majestic northern neighbour, Mount Rainier (4391 m), has more than 90 square kilometres of glaciers, including Emmons Glacier (11 sq km), the largest in the United States outside Alaska. Because of the erosive power of its large glaciers Mount Rainier's cone is less perfect than Mount St Helens' was. Although Mount Rainier has erupted repeatedly in the last thousand years, it has not done so since the mid-1800s. This is just as well as the distribution of lahars formed in the last 10,000 years gives an indication of the devastation that may occur in the event of a major eruption in the future. They range in length from a few kilometres to 110 kilometres (as far as the outskirts of Tacoma), and some fill valleys to a depth of many tens of metres.

Smaller lahars are common even in the absence of an eruption. Heavy rains may initiate jöklhlaups, and as many as nine occurred between 1985 and 1987, the largest in recent years being in October 1947 in Kautz Creek. The large volumes of debris involved have demolished stands of huge Douglas fir trees that cover the flanks of the volcano.

Overleaf Typical effects of glacier modification of the landscape: glacially grooved and striated bedrock, and a few rounded boulders deposited from basal ice. Wordie Gletscher, northern East Greenland.

10

Remaking the Landscape

Glacial Erosional Landforms

The effects of glacial erosion are abundantly clear in many of our most beautiful mountain regions, although now devoid of glaciers, such as the Rocky Mountains of the USA or the highland areas of the British Isles. Sharp peaks and steep-sided, flat-bottomed valleys are typical manifestations of the effectiveness of glacial erosion.

These mountain regions also possess a variety of heaps of sand, gravel and mixed sediments which are the product of glacial deposition. Even more abundant glacial deposits are found in the lowland regions bordering mountain ranges, or on the plains that underlay the outer limits of the last great ice sheets.

The effects of erosion can be seen on all scales – from small outcrops of bedrock to the world's highest peaks, and to the vast areas of scoured low rocky country of the Canadian Shield. The distinctive imprint left by glaciers permits us to recognize the effects of glaciation in areas that have not been covered by ice for many thousands or even millions of years.

Small-scale features

On the smallest scale of centimetres to metres, glacial erosion is represented by striated, polished and grooved rock surfaces, features which are the result of debris-laden ice at the glacier sole sliding over a slab of rock. Associated with them are smaller features such as 'chattermarks' and 'crescentic gouges' which are the result of a repeated juddering of an ice-embedded stone on the rock surface. In addition downstream of a bump in the bed the glacier plucks at the bedrock, creating a jagged rockface. Other features, eroded with the help of meltwater under pressure, take the form of small, irregular, smooth hollows in bedrock, known as plastically moulded forms or 'p-forms'.

Intermediate-scale features

The smaller features are often mirrored on a scale of tens to hundreds of metres. For example, large grooves in areas of scoured low relief that are many times longer than they are wide resemble striations in form. Other larger features include rocky knolls that are convex and smooth

Above The rock over which ice moves is subject to polishing, scratching and plucking. This abraded surface in Varangerfjord, northern Norway, indicates ice movement from right to left. The polish and striations stand out on the wet surface. Towards the right is a crescentic gouge; the down-glacier face of the gouge is always steepest, even if the crescent is orientated the other way.

Above right Smoothed bedrock on the western Scottish islands of the Garvellachs that was grooved by ice and water under high pressure during the last glaciation, within the last 80,000 years. This is one example of an extensive family of features called 'p-forms'.

Right In this view of the Tuolomne Meadows in California's Yosemite National Park, a large outcrop of bare granite, the Lembert Dome, is prominent. This is an exceptionally large *roche moutonnée* and clearly shows that the ice moved from the distant right to the near left.

but striated on their upstream side, and irregular and plucked on the slope facing downstream. These knolls, which are known as *roches moutonnées* after a wavy French wig popular in the eighteenth century, are common at the bottom and along the flanks of glaciated valleys. Sometimes deposition occurs in the lee of a rock outcrop, and the result is a 'crag-and-tail' feature.

Large-scale features

The larger-scale landforms in mountain regions are the most impressive features of glacial activity, as in the glacial erosional landscapes of the Scottish Highlands, the English Lake District and Snowdonia in Wales; parts of the Rocky Mountains and Sierra Nevada in North America; the Pyrenees, much of Scandinavia, and the Alps in continental Europe; and the Urals and ranges in Siberia and Japan in Asia. In low latitudes, abundant forms occur in the Himalayan and associated mountains, and even near the summits of the highest mountains of Africa and New Guinea. Straddling both hemispheres, the Andes also have large-scale glacial landforms, as do New Zealand's Southern Alps and the Tasmanian highlands in Australia.

Cirques, arêtes and horns

In mountain areas without extreme contrasts of height, rocky hollows can often be found that have been excavated out of the higher parts of the mountains by small glaciers that have eroded backwards and downwards. These hollows are known internationally as *cirques* (from the French), though terms like *corrie* (Scottish) and *cwm* (Welsh, pronounced koom) are also widely used in Britain. They typically measure a few kilometres in length and width, and about half to a third of their length in height, but some such as the Walcott Cirque in Antarctica, measures tens of kilometres in length. But whether it is a large one, or a moderate one like the Western Cwm on Mount Everest, or small, the length-to-length height ratio is much the same. All cirques have a steep, frequently vertical headwall, and many have overdeepened basins with small lakes or 'tarns'. In the steepest alpine terrain, however, headwall and downward erosion has normally not been sufficient to create such lakes and the cirque floors slope outwards.

Much of the North-west Highlands of Scotland reveals crystalline Lewisian rocks over 2000 million years old. These older rocks have been modified by ice, and the series of knolls (*roches moutonnées*) shows clear evidence of ice movement from right to left. In the background, the 800 million year old Torridonian sandstone mass of An Teallach (1062 m) survived the onslaught of ice and carried large cirque glaciers of its own. Also visible here is an important indirect effect of glaciation, the rebound of the earth's crust after the ice load was removed to create a raised beach. Such raised beaches are common throughout northern Britain.

Small armchair-shaped hollows on a mountainside which have been scoured out by small glaciers are known as *cirques* or corries. Many cirques contain overdeepened basins with small lakes called tarns. A moraine dam may also occur near the lip of the cirque. Cwm Glas in the Snowdon range of North Wales (*cwm* is Welsh for a hollow).

Close to sea level, in north-western Spitsbergen, this cirque, with its steep sides and near-vertical back wall, is a particularly fine example.

Narrow ridges, often with pinnacles, left after the back walls of adjacent cirques have met through glacial erosion, are called *arêtes*. A well-known one in Britain is Striding Edge on Helvellyn, the mountain in the background (950 m). There are well-developed cirques on both sides of the arête, which were last occupied by ice about 10,000 years ago.

Glaciated valleys have a distinctive form which distinguishes them from river valleys: they are U-shaped, with steep sides and flat bottoms. Many highland areas, although now far-removed from glaciers, have such features. Glen Rosa on the Isle of Arran in south-western Scotland is typical: the slopes are smooth and the spurs that would have characterized the former river valley have been truncated. At the head of the valley other glacial erosional features can be seen, including cirques, the horn of Cir Mhor (799 m) and an ice-eroded col (to the right of the main peak). Arran was last occupied by glaciers 10,000 years ago.

Lauterbrunnental in the Bernese Oberland is often cited as the textbook example of a glaciated valley, on account of its near-vertical valley sides and well-defined U-shape . However, the valley, carved out principally by glaciers descending the slopes of the Breithorn (in the background) and Jungfrau (to the left), is rather an extreme example of erosion; normally the valley sides are less steep.

If cirque glaciers on opposite sides of a mountain erode backwards sufficiently they may meet and a steep-sided rock ridge develops, known as an *arête* (from the French) or *Grat* (German). In Britain arêtes such as Crib Goch in Wales, Striding Edge in the Lake District, Aonach Eagach and the Cuillin ridge in Scotland provide popular scrambling or climbing routes, and in the Alps or western Cordillera of North America, they provide aesthetically pleasing and challenging routes to the summits. Many of the first ascents of Alpine peaks, such as the Matterhorn and Weisshorn, were made by way of arêtes.

Where three or more cirque glaciers have eroded backwards to the extent that the highest ground is no longer immune from erosion, a sharp, pointed peak or 'horn' is produced, with a series of arêtes between each pair of glaciers rising steeply up to the summit. The best known of all horns is the Matterhorn on the Swiss–Italian border, but other well-known and well-formed horns are Mount Assiniboine in the Canadian Rockies, Mount Aspiring in New Zealand's Southern Alps and K2 in the Karakorum. In Britain, one of the best examples is Cir Mhor on the Isle of Arran.

Glaciated valleys

Glaciated valleys that were formerly occupied by the main glaciers descending from the higher mountains have a characteristic form. In cross-section they are approximately U-shaped, although their sides are only rarely truly vertical – Lauterbrunnental in the Bernese Alps and the Yosemite Valley in California are famous examples of valleys which do have near-vertical walls, while the Scottish Highlands, of which Glencoe in the Grampians is perhaps the finest, provide good examples of less steeply sided glacial valleys.

In many glacial valleys enhanced erosion has been such as to create a series of rock basins, filled with lakes, which are formed when a tributary glacier joins the main one. Later these lakes will be partly or totally filled with sediment. Many glacial valleys have a stepped longitudinal profile, or a series of rock barriers (*riegels*) extending part or the whole way across the valley.

The level to which a valley glacier has eroded is clearly marked by a change in slope, as well as by the preservation of spurs above the glacier

Yosemite Valley, California. A hanging valley is perched high above the floor of the main valley, linked by the Bridal Veil Falls which tumble down the vertical rock face, staining it black.

level, as up to the erosion level the side of a glaciated valley normally consists of a single clean sweep, the spurs of the mountains that would have existed in preglacial times having been truncated. Some tributary glaciers may not have had the erosive power to cut down to the level of the main valley, and hanging valleys result, often with fine waterfalls descending from them today. In other cases, valley glaciers may have spilled over the low saddles or cols of the bounding ridges into another valley, resulting in 'breached watersheds'.

Fjords

Fjords are flooded overdeepened glaciated valleys, formed because valley glaciers erode deeply in their middle reaches and, if extending into the sea, to a depth well below sea level. The term *fjord* is of Norwegian origin (spelt *fiord* in North America), and it is Norway which has one of the finest fjord coastlines in the world. The whole western coast of the country is indented with them, the longest and deepest being Sognefjord (200 km long and 1300 m deep). But many other moderate- to high-latitude countries have fjords too. Those in western Scotland are known as *lochs* (as are the glaciated valley lakes), but are small by world standards. Greenland, where they are still being formed, has the world's longest fjords, with the combined Nordvestfjord and Scoresby Sund on the east coast measuring 350 kilometres. In the Americas well-developed fjord coastlines occur in southern Alaska, British Columbia and southern Chile, many of them also still influenced by glaciers. The islands of the High Arctic have many smaller fjords, while on the other side of the world south-western New Zealand has a well-developed fjord coastline as impressive as any, and the Antarctic Peninsula has many fjords that are filled by glaciers. The subantarctic island of South Georgia has numerous short fjords still occupied by glaciers.

Like glaciated valleys fjords may comprise several rock basins, but the simplest of them are deepest at their heads and become gradually more shallow towards the sea. Many have a shallow or even partly exposed sill at their seaward limits, and most have near-vertical rock walls. Like valleys, they are approximately U-shaped in cross-section, and with time they also become filled by sediment. Hanging valleys and waterfalls are common features.

Scoured bedrock of low relief

The vast areas of the gently undulating Canadian Shield contain large-scale glacial erosional landforms of a scenically less dramatic nature. This area of hard crystalline, Precambrian rock bears evidence of erosion along the structural grain of the bedrock, with long linear erosional forms, in many cases containing elongate lakes or boggy hollows. (A similar effect is apparent on parts of the Baltic Shield of Scandinavia.) Parts of north-western Scotland show the same process short of completion, where isolated sandstone peaks stand proud above the heavily scoured ancient crystalline rocks with their little knolls and small lakes; this is called 'knock and lochan' topography.

Glacial depositional landforms

'Towards the end of seven days the waters of the flood came upon the earth. In the year when Noah was six hundred years old, on the seventeenth day of the second month, on that very day, all the springs of the great abyss broke through, the windows of the sky were opened, and rain fell on the earth for forty days and forty nights. . . . More and more the waters increased over the earth until they covered all the high mountains everywhere under heaven. The waters increased and the mountains were covered to a depth of fifteen cubits. Every living creature that moves on the earth perished, birds, cattle, wild animals, all reptiles, and all mankind . . . only Noah and his company in the ark survived' (*Genesis* ch. 7, vv. 10–12, 20–21, 23).

This was the explanation widely held in the early nineteenth century to explain the distribution of what we now know to be glacial deposits over northern Europe. At the time controversy raged as to whether or not the Bible should be taken literally, and this particularly affected geologists. It was recognized that many large blocks of rock, now called erratics, scattered over much of northern Europe had been transported great distances. The influence of glaciers was not then known, and the geologists of the day could only conceive that the blocks were moved during catastrophic floods. Thus Noah's Flood was given as an explanation for these deposits, which were given the name *Diluvium* (the Latin word for flood or deluge).

The north-west face of the Bietschhorn (3934 m) in the Bernese Oberland of Switzerland. This typical horn is characterized by the fine curving arête in the centre and left.

Previous page Fjords, glaciated valleys flooded by the sea, are common in several temperate to high-latitude regions. Among the finest are those in the Fiordland National Park of New Zealand, of which Milford Sound, dominated by the cloud-capped Mitre Peak (1692 m), is the best known. Dense rain forest covers its rocky slopes, and the overcast sky and light rain create an austere atmosphere.

In fact, the idea that glaciers could be responsible for transporting large boulders was already gaining currency, following the work of Swiss naturalists in the Alps in the late eighteenth and early nineteenth centuries, but the widespread acceptance and application of the theory only grew very slowly.

Glacial and meltwater deposits are the components of a wide range of interesting landforms, which, although less dramatic than erosional ones, are nevertheless distinctive features of the environment. The rolling countryside with fertile fields and wooded knolls, so characteristic of the lowlands of central and northern Europe and the northern Mid-West of the USA, are dominated by such features. Yet, although best developed in the lowland areas that once were covered by continental ice sheets, depositional landforms are also present in highland areas where glacial erosion was dominant.

Moraines

The most easily identifiable of depositional landforms are moraines, since they are frequently long, sharp-crested ridges or ridge complexes, made up of a mixture of till and other deposits pushed up during glacial advances. In a valley setting, the furthest advance of a glacier is marked by a terminal moraine, normally arcuate in form, reflecting the original shape of the glacier snout. Such features usually range from a few to fifty or more metres in height, but because the glacier's retreat was normally accompanied by powerful streams, only remnants may remain. In lowland areas the advances of ice sheets created even larger ridge complexes, that can sometimes be traced for hundreds of kilometres, which were breached in places by huge streams issuing from the retreating ice. Lakes frequently formed behind these terminal moraine complexes, the Great Lakes of North America and Zürichsee in Switzerland being prominent examples.

Lateral moraines appeared along the sides of valleys, but these have a much poorer chance of surviving; hillslope movement and subsequent rockfalls combine to obliterate them. However, in areas of recent retreat, unstable ridges dating back to the 'Little Ice Age' of around 1650–1850 are still in evidence high above the present glacier tongue and down-valley of the glacier's snout. The ridge commonly stands out

Vadret da Tschierva in the Bernina mountains of Switzerland displays a pair of fine lateral moraines. These knife-sharp features were deposited during the Little Ice Age of the seventeenth to early-nineteenth centuries. The glacier is once again advancing dramatically in this photograph of 1986, refilling the space between the moraines as the steep convex snout moves downslope. The highest peak at the head of the glacier is Piz Bernina (4020 m).

Above A 60-tonne erratic boulder east of Zürich in the Swiss Mittelland lies 60 kilometres from its place of origin in the Alpine valley of Linthal.

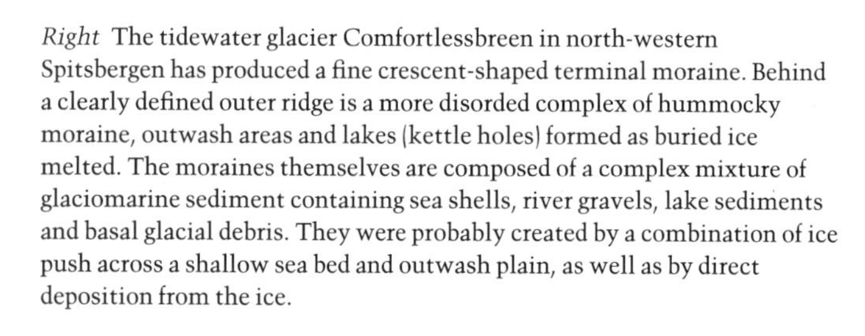

Right The tidewater glacier Comfortlessbreen in north-western Spitsbergen has produced a fine crescent-shaped terminal moraine. Behind a clearly defined outer ridge is a more disorded complex of hummocky moraine, outwash areas and lakes (kettle holes) formed as buried ice melted. The moraines themselves are composed of a complex mixture of glaciomarine sediment containing sea shells, river gravels, lake sediments and basal glacial debris. They were probably created by a combination of ice push across a shallow sea bed and outwash plain, as well as by direct deposition from the ice.

from the valley sides, creating a subsidiary valley, with its own stream and ponds, between the ridge and the valley side, well above the floor of the main valley. A steep face of bouldery till, held temporarily together by the clay in the deposit, commonly formed on the glacier side of such moraines, and today rainwater creates a furrowed pattern down them, making scrambling up or down them surprisingly difficult. In contrast, the side of the moraine nearer the hillside tends to be well covered, plant growth having been possible even when the ice was in position.

Once the ice has gone, these ridges collapse rapidly. Where a valley opens out, lateral moraines may leave the valley sides and merge with terminal moraines. Between them and the retreating glacier irregular wastage of stagnant ice creates a random assemblage of hillocks and hollows filled with water. These hummocky moraines are common in glaciated valleys, and are a good clue as to how the ice retreated. However, not all valleys have them.

Each of the above types of moraine may have a core of ice, and the volume of debris may be small; dark wet patches of unstable debris are clear indications of an ice core. These ice-cored moraines may survive for many decades or even centuries.

There are other types of moraine up to a few metres high associated with existing glaciers, but they rarely survive for more than a few decades. Small annual 'push moraines' may form as a glacier advances a few metres in winter when ablation largely ceases, and a whole series of such parallel ridges may develop if the longer term trend is one of retreat. Sometimes, ice movement across a plain of till may generate 'fluted moraines' – long, straight, parallel, smoothly rounded ridges parallel to the ice flow direction, which may be exposed if the ice front retreats steadily without *in situ* stagnation.

A special set of moraines is associated with surge-type glaciers, that is glaciers that periodically advance catastrophically, as described in chapter 5. Debris may be thrust up from the glacier bed during a surge to form curving ridges a few metres high parallel to the snout. At the end of the surge wholesale stagnation of ice occurs, creating a hummocky and pitted morainic topography on which the thrust ridges have been let down. Discriminating between non-surge and surge-type moraines is, in fact, an important way of assessing whether a particular glacier advance was the result of climatic change, or of periodic instability of the ice mass.

If a glacier or ice sheet rides over a plain of till or heaps of moraines, the material becomes quite gooey and is easily moulded into new shapes or eroded. One product of this process is the generation of 'drumlins', streamlined hillocks of till, sometimes draped over bedrock knolls. A drumlin is shaped like an inverted spoon, with the steep slope in the upstream direction, and is orientated parallel to the ice flow direction. They reach 100 metres or more in length and are up to 50 metres high, often covering large areas in a type of landscape known as basket-of-eggs topography. The Eden Valley of north-west England is a good example, and extensive drumlin fields also occur in New England.

Although glacial transport is best demonstrated by the above landforms, isolated erratics are also good indicators of the former extent of ice and the directions of flow. Large blocks that fell on a glacier surface, or boulders in the bed of a glacier, may end up hundreds of kilometres from their place of origin.

In some cases moraines may contain valuable minerals, and even if the moraine itself is uneconomic, by reconstructing flow paths back to the source rocks it is possible to locate viable mineral reserves. This method of mineral location has been used most widely in Canada and Finland.

The power of glacial meltwater

Meltwater within, at the side, or beyond the limits of a glacier also creates landforms. Wide fluctuations in discharge between summer and winter create unstable 'braided' stream channel systems, where the streams modify, sort and redistribute glacial debris creating 'outwash plains', both beyond the terminal moraine system and directly in front of the retreating glacier. Such systems normally extend across the entire width of a flat-bottomed valley, as can be seen in some fine examples in New Zealand and Alaska. In lowland areas these plains may be many tens of kilometres wide. They are known as *sandar* (singular *sandur*), a term from Iceland where they extend along much of the south coast, south of the ice cap of Vatnajökull.

Glacial outwash deposits often bury remnants of dead glacier ice. As the ice slowly melts, steep-sided, water-filled hollows, known as kettles

Left Hummocky moraine is characteristic of areas of stagnating or slowly retreating ice, as in front of Midre Lövenbreen, Kongsfjorden, Spitsbergen.

Above Drumlins are streamlined forms resulting from a combination of erosional and depositional processes beneath moving ice. Often they occur in groups as here in the Hirzel area south of Zürich, where two large Ice Age glaciers coallesced. The fertile morainic ground in this area is now an important source for some of Zürich's drinking water.

The wide fluctuations in the discharge of glacial rivers create an unstable network of braided channels. Here we are looking down one of the outlet streams from Casement Glacier, Glacier Bay, Alaska, at a time of moderately high discharge in June 1986. The small, roughly circular ponds in the outwash gravels are kettle holes, formed as buried ice melts. The glacier itself on the left is totally covered by debris.

or kettleholes (a term of Celtic origin) develop. Beneath the glacier itself the channels of water become choked with debris. As the ice melts, long, narrow ridges of sand and gravel may be left standing above the general level of the outwash plain, in features known as *eskers* (another Celtic term). Varying in height from a few to tens of metres, they wind for hundreds of metres across the landscape. In Finland, which is particularly noted for them, they provide convenient flat-topped ridges on which roads are often built through lake-studded country.

Stream deposits frequently accumulate adjacent to the ice, which are left as isolated hillocks when the ice melts. The term *kame* (an old Scottish word) is used to describe ice-contact meltwater deposits formed in this way parallel to the ice front, and kame terraces are formed at the side of a glacier, especially where a tributary stream enters the main valley. As the ice melts, a level, gently sloping platform is left, perched on the hillside. In addition to stream deposits, a kame and kame terrace may also be composed of lake sediments.

Apart from their character as distinctive elements of a glacially influenced landscape, outwash, esker and kame deposits provide much of the sand and gravel needed for road and building construction around the temperate regions of the world, especially in Europe and North America.

11
Wildlife

Surprising as it may seem, glaciers and their surroundings are often havens for wild life. Camping on a glacier a thousand metres up amongst the icefields of Spitsbergen the authors have been disturbed as they tried to sleep by the constant chatter and chuckling of a colony of fulmars, nesting on bare rock cliffs nearby, although 30 kilometres from the coast. We heard the cry of an Arctic fox, and looked out to see that elegant, white-coated animal trotting back and forth below the cliffs, eyeing the nesting birds with eager anticipation, and obviously waiting for an unfortunate chick to fall out of its nest.

Glaciers are not totally lifeless, despite the harshness of the environment, but around the world, many different animals and plants live and die on and around them, uniquely adapted to the cold. Some species live entirely in the main glacierized regions while others make their homes around mountain glaciers, where animals can usually move down to lower ground, if conditions become too severe.

Antarctica

The most hostile environment on Earth is Antarctica. In reality, the region is a cold desert, characterized by low snowfall, lack of water, exposure to the wind, and soils that lack organic matter. The number of species capable of living under these conditions is small, even in and around the seas bordering the continent where the actual numbers of animals is vast, especially of the twelve species of birds and four of seals that breed there. The low diversity of species reflects the continent's isolation from the nearest land, from which it is separated by 650 kilometres of stormy seas that have discouraged migration.

Land animals in Antarctica are small; indeed, there is nothing bigger than a horsefly, with the mite being the most common creature living entirely on the land. In contrast, the sea is home to a number of large, warm-blooded creatures which are top of the food chain, for example the world's largest seabirds, seals and whales. One of its residents, the blue whale, is the largest animal the world has ever known, growing fat on the abundance of food, especially krill, that the sea provides. Unfortunately, man has ruthlessly hunted this beast to the extent that it now needs total protection.

Previous page Emperor penguins at the edge of the fast ice in McMurdo Sound, Antarctica. In the background is the glacier-covered northern tip of Ross Island, a breeding ground for these flightless birds.

Most marine birds and mammals spend more of their time at sea than on the ice, and feed entirely in the water. However, some sea birds roost and breed far inland – for example, snow petrels and skuas have been found in small colonies 2,000 metres high on the nunataks of Dronning Maud Land, 300 kilometres from the sea. Skuas venture even further into the icy interior of Antarctica, and the ill-fated polar party of Captain R. F. Scott in 1912 recorded one only about 250 kilometres from the South Pole, over 800 kilometres from the nearest open water.

To many people, the most fascinating of all birds is the penguin. Two species, the emperor and the Adélie, breed on the continent, especially on stable sea ice, or on ledges or in caves in ice cliffs and ice shelves, and even more breed on the more northerly offshore islands. More than any other bird, the emperor penguin is uniquely adapted to the frigid climate, although its population is estimated to be only about a quarter of a million. It lays its eggs in early winter and incubates them through the coldest months of the year. The males take over from the females shortly after the eggs have been laid, huddling together against the fierce blizzards to conserve heat, with the eggs on their feet, covered by a fold of feathered skin. After about sixty-five days, at about the time that hatching takes place, the by now well-fed females return, allowing the hungry male to plod north across the sea ice to find food in open water.

The emperor's breeding habits have long been a source of fascination to biologists, ever since Edward Wilson, chief scientist on Scott's last expedition, together with 'Birdie' Bowers and Apsley Cherry-Garrard, in 1911 made one of the most arduous journeys ever undertaken in the cause of science. Graphically described by Cherry-Garrard in *The Worst Journey in the World*, this expedition was undertaken in the darkness of the depths of winter in order to capture an emperor penguin egg soon after it was laid, in the hope that the primitive embryo would establish a link between reptilian scales and feathers, and thus throw light on the origin of all birds. The privations this party suffered give a good indication of the conditions experienced by those emperor penguins that remain at the glacier edge in winter. Pulling sledges weighing 343 kilograms and containing food and equipment for six weeks, the three men had to cross part of the Ross Ice Shelf, where the temperature plummeted to −61°C (−77°F) and often failed to reach −50°C (−60°F) all day.

A family of arctic hares grazes peacefully on the shores of Antarctic Sund in East Greenland, as icebergs drift by in the background.

Pulling the sledges was made difficult by the low temperatures which gave the snow a sandy texture, and they had to do it in relay. They also had to wend their way in the dark through pressure ridges and crevasses where the ice shelf piled up against Ross Island.

Eventually, the party reached Cape Crozier, and with difficulty scrambled down steep icy slopes to reach what was then the only known emperor penguin colony. After collecting a number of eggs, the men established a camp above the colony, only to have it swept away during a violent blizzard. For three days they huddled under a groundsheet ensconced in their iced-up sleeping bags, while the blizzard shrieked over their heads. Fortunately, they found the tent undamaged, for without it their survival would have been in doubt. Exhausted by the hardships of the preceding four weeks, the party now had to retrace its

Adélie penguins on the Ekstroem Ice Shelf, near the German research station of Georg-von-Neumayer in East Antarctica.

This Weddell seal has swum 40 kilometres beneath the sea ice to a crack adjacent to the Bowers Piedmont Glacier, McMurdo Sound, Antarctica, and given birth to her single pup.

steps to the expedition base, nearly 100 kilometres away. Despite falling asleep on the march, they eventually reached safety, thirty-six days later, after enduring probably the harshest conditions ever to be survived by men. Sadly, both Wilson and Bowers lost their lives later in the expedition, after reaching the South Pole with Scott.

Penguins and seals probably spend half their lives in the waters surrounding Antarctica, which for most of the year are warmer than the air. Warm-blooded animals are well adapted to take advantage of the rich food resources in the sea. Heat loss is reduced to a minimum by dense, water-repellant plumage in the penguin and petrel, and by the thick, tough skin of the whale and seal, and a thick layer of subcutaneous fat or blubber provides additional insulation. A further adaptive feature is their compactness, the extremities of seals and penguins are generally short and bony, so little blood needs to circulate through exposed areas. Their large size is also an advantage in the cold climate, since a low ratio of surface area to volume conserves heat more efficiently than the large ratio of a small animal.

Although seals and birds can tolerate very low temperatures in still air, they prefer to be in water when it is windy. However, if they have to, such as when they are protecting their young, they are capable of remaining at their post without harm in blizzards that last several days. The young may not be so fortunate, and in severe storms the mortality amongst juveniles is high.

Some animals do curious things in Antarctica. The strangest is the behaviour of crabeater seals in crawling inland over rocky debris or up glaciers for many kilometres to die. Dried out carcasses have been seen as much as 750 metres above sea level and 70 kilometres from the coast.

Antarctica had a lush cover of vegetation, including forests, until the ice sheet developed about 36 million years ago, but it was then largely wiped out. However, even today it is not as utterly devoid of vegetation as might appear at a first glance, with various species of algae, lichens, mosses and fungi. The greatest diversity occurs in the warmer maritime parts of the Antarctic Peninsula and offshore islands, where even a couple of species of flowering plants also occur. Mosses and lichens have been recorded on nunataks as far south as latitudes 84° and 86° respectively.

An Arctic fox shares the same drinking water as glaciologists on Axel Heiberg Island in the Canadian Arctic.

When ice retreats the first plants to colonize are algae and soil micro-organisms, with mosses and lichens appearing later, as can be observed both on morainic debris and bare rocks. They all grow best along the emphemeral water courses and around snow-banks. Antarctic plants survive because they have developed a strong resistance to frost and drying out, and have acquired the ability to grow rapidly in the brief periods when conditions are favourable.

The Arctic

The polar desert environment that characterizes much of Antarctica is found in a few inland parts of the high Arctic, but is less impoverished. Indeed, most parts of the Arctic have a much more tolerable climate, so biologically the region differs greatly from Antarctica. Although extensive glacierized areas do exist in the Arctic, ice-free areas are much more widespread. Furthermore, parts of the Arctic remained ice-free even during the ice ages, allowing evolution to proceed without interruption. Thus, there is a much greater diversity of plant and animal species than in the south polar region.

The most striking difference is the presence of large land mammals that roam freely across snow and icefields as well as the tundra – the sparsely vegetated ground that is underlaid by a thick layer of permanently frozen ground called permafrost. A total of forty-eight species of land mammals are found in the Arctic, comprising shrews, hares, rodents, wolves, foxes, bears and deer, but, although more than Antarctica, even this diversity is low compared with more temperate regions. However, only a few of these live in proximity to the glaciers. Greenland, for example, has only nine species: Arctic hare, Arctic lemming, grey wolf, Arctic fox, polar bear, ermine, wolverine, caribou and musk ox.

Arctic animals illustrate a variety of adaptations to the hostile winter conditions. Some avoid the severity of winter by migrating south; for example, the majority of birds, and many of the herds of caribou in North America. On the other hand, the reindeer on Svalbard have nowhere to go, and seem to thrive on the cold: you often see them in summer cooling off on glaciers and snow patches. Other animals go into

Left The Arctic poppy (*Papaver alpinum*), one of the earliest flowers to colonize ground vacated by ice in the Expedition River area of Axel Heiberg Island, Canada.

Right and below The ptarmigan is a resident bird of Arctic and northern mountain regions, photographed here on Axel Heiberg Island, Canadian Arctic. Its winter plumage (white) and summer plumage (brown) show how its camouflage adapts to the changing seasons.

Solitary male musk ox on Ymer Ø, East Greenland, shortly before taking offence at being photographed and charging the cameraman.

Caribou foraging for lichens in front of Thompson Glacier and its push moraine, Axel Heiberg Island, Canadian Arctic.

hibernation, such as the Alaskan marmot and Arctic ground squirrel. Although bears strictly do not hibernate, the grizzly, which is found around the glaciers of Alaska, enjoys a long period of sleep in winter.

The Arctic fox is adapted in several interesting ways. It remains active all the year round, in autumn switching from a brown coat to a dense white coat to match the snow of winter, and has furry paws to insulate its feet from the frozen ground. It is able to maintain a steady heat production, and only needs to increase it when the temperature declines to −40°C. The Arctic fox ranges widely through the Arctic, venturing on to glaciers as mentioned above, and in winter on to the sea ice, and has been sighted within 140 kilometres of the North Pole, 80 kilometres from the nearest land. The fox lives off birds and lemmings, and scavenges off the remains of seals killed by polar bears.

The carnivorous grey wolf is a survivor from the ice age that spread through much of the Arctic when the ice retreated. Hunting in packs, it feeds principally off sick and weak caribou in Canada and Alaska, but in Greenland relies on small animals.

Like the fox and wolf, the musk ox remains active throughout the year, favouring dry areas where the snowfall is low, so that it can scrape away at mosses and lichens. It carries a thick, winter coat that is softer than normal wool. In summer the animal moults, and acquires a very untidy appearance, as it gradually loses trains of knotted wool.

The Arctic hare differs from more southerly hares in having a much thicker insulation. In winter it turns white, and in more northerly parts remains white all year round. Unlike most species of hare it sometimes forms large herds, but mostly can be seen in groups of half a dozen or so.

Small animals, such as rodents, cannot compensate for heat loss by improving their insulation, so instead they increase their metabolic rate. They also use the insulation provided by their surroundings – by sheltering beneath the snow cover.

The world's largest carnivore, the polar bear, epitomizes the animal world's adaptation to the harsh Arctic environment. Polar bears are widely, but thinly, distributed over the Arctic sea ice, but they breed in onshore dens, usually in snowbanks on barren coastal hillsides. They range over hundreds of kilometres, right across the frozen Arctic Ocean. Although they rarely venture on to glaciers, they have been known to

Right The Arctic willow, flowering here in East Greenland, is one of the main shrubs of the high Arctic. Although often only a few centimetres high specimens may be a hundred years old or more, like their much bigger relatives in temperate regions.

Far right Yellow mountain saxifrage (*Saxifraga aizoides*) in East Greenland.

Below right Broad-leaved willow-herb (*Epilobium montanum*), East Greenland.

cross icefields, such as on Svalbard when rapid break up of the sea ice on the west coast makes it impossible for the bear to catch seals, forcing it to cross the land ice to reach the more permanent sea ice on the east coast.

Few species of bird remain on the Arctic tundra during the winter, and those that do, such as the ptarmigan, have adapted by reducing their metabolic rate at low temperatures and increasing the density of their feathers. However, in summer the Arctic supports nearly 200 species of bird, which have migrated both overland and over the sea. They arrive in springtime as the snow begins to melt, plants begin to flower and the days become long. One of the most impressive sights in nature is the thousands of seabirds nesting on, and wheeling about the steep cliffs of a fjord, with a glacier calving at its head, producing small icebergs which provide a perch for the birds, and also for seals. Although most birds prefer to nest on coastal cliffs, some sea birds prefer nunataks amongst the highland icefields, so few parts of the Arctic are devoid of life. Some breeding birds cover enormous distances, but none can compete with the Arctic Tern which nests in the Arctic and migrates to the Antarctic for the southern hemisphere summer.

In terms of numbers, insects are the most abundant of Arctic creatures. They play a vital role in the ecosystem, providing an important source of food for many birds. Most memorable for the human are the mosquitoes and midges which inflict so much misery on travellers when the weather is warm, but there are many other less noticeable species.

Arctic plants have also adapted to the severe climate in several ways. Most obvious is their smallness of size, which minimizes desiccation from winds and allows them to nestle under an insulating cover of snow. Temperatures are also higher at ground level during the growing season, facilitating flower and seed production. Among their numerous other adaptations, a striking one is the early development of buds, which allows them to flower at the earliest opportunity in the spring. Their dark-coloured leaves and stems are also noticeable; this allows a greater heat absorption, especially if the leaf cover is dense, as in the heathers and saxifrages which protect the plant from abrasion by snow.

Arctic plants have to be further adapted to the seasonal variations in

Above Stone ibex silhouetted against the glaciers of Fiescherhorn (4025 m) in the Bernese Oberland of Switzerland.

Right Mountain goats on steep, unstable, rocky terrain above Variegated Glacier in Alaska.

light intensity, such as an ability to vary their rates of photosynthesis. They also can withstand total freeze-up of the ground for eight to ten months of the year, as well as summer frosts.

Alpine regions

In the mountainous areas of more temperate latitudes glaciers frequently descend into zones where the climate is much less severe than in the polar regions. Thus, when the ice retreats, plants may colonize the newly exposed ground at an amazing speed, enjoying the fertility that is often provided by sediments rich in minerals. The so-called pioneer plants are the first to arrive, having been transported in seed form by the wind and birds and then germinating in little more than sand and clay. These early plants include saxifrages and grasses, and may often be found within two years of the ice retreating. Decaying organic material from these pioneer plants then prepares the ground for more demanding species, such as the larger flowering plants, *Alpenrosen* and asters, as well as shrubs of alder and willow. The colonization sequence is complete when coniferous trees have taken over, cutting out much of the sunlight required by the pioneer plants.

A marmot, the variously inquisitive or shy rodent that inhabits high Alpine pastures. This one, on the lateral moraine of Gornergletscher in Switzerland, was distinctively of the former disposition, and begged for biscuits!

Left Cotton grass (*Eriophorum angustifolium*) growing in a boggy area near the snout of Sheridan Glacier, Copper River area, Alaska.

Right King of the Alps, Pontresina, Switzerland.

Right Purple saxifrage (*Saxifraga oppositifolia*), Pontresina, Switzerland.

Spring Gentian (*Gentiana verna*), an evergreen perennial that produces bright blue flowers in early spring. Griesgletscher, Switzerland.

Grasses and flowering plants sometimes grow on the surface of Alpine glaciers if the debris cover is thick and the ice inactive, while in Alaska and Patagonia even trees grow on debris covering the snouts of stagnant glaciers.

The vegetation provides food for grazing animals. In the Alps the stone ibex and chamois are common, scrambling with ease over precipitous rocky hillsides, while marmots emerging from holes in the ground make a whistling sound to warn their mates at the first signs of danger. Above soar birds of prey, such as eagles, waiting for the opportunity to swoop on unsuspecting rodents below.

In the mountains of western North America, Dahl sheep and mountain goats inhabit terrain similar to that of the ibex and chamois, whilst in less rugged terrain moose, elk, bears and other animals frequently forage for food.

Scandinavia is well known for its lemming years, when a population explosion creates the alleged suicidal tendency of the rodents to drown themselves in the sea. Less widely reported is that many of them, as we ourselves have seen, perish running up on to glaciers and dying of cold.

Most primitive of all these life forms is red algae which in summer frequently stain the snow pink. There is even a creature called an ice worm in North America which completes its life cycle on the snow and ice. A small, wingless insect, the glacier flea, also lives exclusively on glaciers in the Alps and North America, living off wind-blown pollen.

The Himalayas are known for the reputed occurrence of a mysterious animal, the Yeti or Abominable Snowman, but except for its tracks, it has proved elusive to western mountaineers. Some Sherpas talk of a hairy, monkey-like animal, but others think the creature is more likely to be a bear. Whatever the truth, it seems that some unidentified animal prowls around the glaciers of the Himalayas, and rarely comes down to lower altitudes.

Thus from the minute to the mythical, glacier country demonstrates the animal kingdom at its most resilient, and those people who venture into such areas may be privileged to see a remarkable variety of life, well-adapted to the hostile environment.

12

Ice, Climate and Civilization

The study of glaciers is important for many reasons. Perhaps most fundamentally, glaciers and ice sheets respond to climatic change. Thus ice cores and glacial sediment cores can provide unequalled records of climatic evolution both within the span of a human being's lifetime, and on longer time scales, including those of millions of years. If we can establish the environmental record of the past, we will be better placed to understand the causes of climatic and sea level changes, and be in a sounder position for working out what may happen in the future.

Dangerous glaciers impinge directly on the lives of people in mountain regions, and some have been responsible for huge loss of life. Again we need to understand them better if future catastrophes are to be averted. Despite this, glaciers provide considerable benefits to human society, notably in terms of providing a reliable water supply and energy resources, and not least they are of considerable value to the tourist industry. The better we understand them, the more useful we are likely to find them.

The frozen climatic record

Tourists who take part in a boat excursion to the calving front of Columbia Glacier in Alaska are often offered a unique drink: a Martini cooled by ice that has calved or been hacked from the glacier. The tour guide will explain that the ice is very old indeed, formed over 10,000 years ago during the last ice age, but any glaciologist present might be tempted to spoil the fun since an active maritime glacier like the Columbia transfers its ice at such an impressive rate that none can be expected to survive for more than a few centuries. However, ice even this young will predate the Industrial Revolution and be fresher than any ice formed naturally since. It is not then surprising that the prospect of finding old ice in more stable glaciers has lured glaciologists and climatologists to the polar ice sheets and ice caps and to high-level ice caps in the temperate and tropical latitudes to seek climatic records. Particularly useful elements and components trapped in the ice include varieties of oxygen, carbon dioxide, and the products of nuclear explosions. These glaciologists have had some remarkable results, obtaining samples that predate the last ice age.

The Weisshorn (4505 m) towers above the village of Randa in the Valais, Switzerland. Several times in the past, parts of the hanging glacier below the summit have broken off and fallen as ice avalanches onto Bisgletscher, the glacier visible in the centre of the photograph. Such ice avalanches are particularly dangerous in winter when they mobilize enormous masses of snow and can reach as far as the village. In 1639, 36 people were killed when this happened.

Geologists have also joined the search for climatic records by drilling into sediments, not only in areas close to present-day ice sheets, but also in the deep ocean basins. These records are much less complete, but go back much further, providing a glacial record that began at least 40 million years ago – much earlier than was once thought possible. The evidence both methods provide of past changes in climate, and the response of the ice sheets and the sea levels, gives us valuable information about how the global system works, and provides an essential database for scientists trying to make predictions about future changes.

Climbers on their way to the Weissmies (4023 m) in the Swiss Alps are dwarfed by the crevasse which they have just crossed. The crevasse wall clearly shows layering as a result of successive snowfall events, with yellow layers marking deposits of dust often carried by winds from the Sahara Desert to central Europe. Dust-falls are extremely useful for dating firn layers, such as those exposed here, or others retrieved by drilling from much greater depths in high-altitude glaciers.

The greenhouse effect

In recent years there has been increasing recognition of the fact that mankind's activities are changing our planet's atmosphere and climate.

The earth benefits naturally from a greenhouse effect whereby carbon dioxide and other gases trap solar radiation and prevent all the sun's heat from bouncing off the surface of the planet and back into space. However, most climatologists now agree that the global temperature is rising, and that this is at least partly due to the increasing levels of various gases, particularly carbon dioxide, in the atmosphere, directly attributable to human activity. These increases arise from a combination of the burning of fossil fuels, which release carbon dioxide, and the destruction of forests, which convert carbon dioxide into oxygen. Increased levels of methane and the depletion of the ozone layer are also acknowledged as having a warming effect on the atmosphere.

Records of past climates

While some climatologists examine the processes at work in the Antarctic atmosphere, others have worked with glaciologists in obtaining deep cores from the sheets below it. They have also taken deep cores from the Greenland ice sheet, as well as shorter ones from ice caps in warmer regions, including the 5650-metre-high Quelccaya ice cap (14°S) in southern Peru, the Dunde ice cap on the Tibetan Plateau (38°N), a highland icefield at 5300 metres on Mount Logan (60°N) in the Canadian St Elias Mountains, and near the summit of the second highest mountain in the Alps (as described below). Drilling operations in these terrains are difficult and costly, but thousands of metres of core have been recovered, which have given us long-term records from most of the main climatic zones on earth.

There are a number of practical reasons why scientists are investigating the atmospheric processes of the distant past. We urgently need accurate information about climatic evolution in order to establish the nature of, and reasons for, climatic changes today, and to predict future trends. It is a matter of great concern that we still cannot say with any certainty why and how ice ages begin and end; nor can we predict how the biggest single controller of rapid sea-level changes, the Antarctic Ice sheet, will respond to climatic warming. However, the climatic record is preserved throughout the entire depth of certain parts of ice sheets and ice caps, where snow and ice accumulation and glacier flow is slow, and melting is negligible. In them relatively thin snow layers are deposited on top of each other annually and remain undisturbed for thousands of years. After retrieving ice cores, the scientist analyses them for air bubbles trapped in the ice which provide us with samples of the atmosphere of the past. For example, the ratios of isotopes (especially heavy and light oxygen and hydrogen/deuterium) in ice indicate air temperatures at the time of deposition of the snow, and carbon dioxide and methane in the bubbles indicate how the burning of fossil fuels has affected atmospheric composition.

A record from central Europe

Most ice-cores have been obtained from sites far distant from industrialized areas. This matters little for gases like carbon dioxide which, given time, will mix more or less evenly throughout the whole troposphere. But pollutants such as sulphur compounds are washed out of the air over what, in global terms, is a relatively small area, so if we want to examine the increase in atmospheric pollution since the industrial revolution we need to obtain ice from as close as possible to the sources of that pollution.

There are a few such suitable sites in the middle of heavily industrialized central Europe, in the Alps, of which Colle Gnifetti at 4452 metres on Monte Rosa on the Swiss–Italian border was chosen for coring. First the cores gave us a clearer understanding of the processes of snow deposition and its transformation to ice at very high altitudes. Accumulation was found to reach a maximum at 3500 metres above sea level, where typically 3 metres of water-equivalent of firn are formed each year, but above this altitude it decreases rapidly. The accumulation at Colle Gnifetti was discovered to take place mainly in summer, rather than in winter, as is the norm of temperate regions. Furthermore, accumulation is more pronounced in relatively warm summers than in cool ones.

The small rate of firn accumulation at high levels on the mountain means that relatively shallow boreholes reach several centuries back in time, an important consideration since drilling is not only costly but also extremely exhausting in the rarefied air near the top of Monte Rosa. On the other hand, the stratigraphic record is incomplete, and in some years all the preceding year's snow, and maybe some from earlier years,

A helicopter has here lifted glaciologists and scientific equipment to the high glacier plateau of Colle Gnifetti on Monte Rosa, where ice cores have yielded much invaluable information about climatic conditions in central Europe during the last 500 years. Atmospheric gases trapped in the ice of Colle Gnifetti document both the pure air of former centuries and the increasingly severe air pollution since the industrial revolution.

is blown off the col. However, Saharan dust storms, nuclear explosions (emitting radioactive isotopes like tritium), and most recently the Chernobyl accident have helped to establish a reasonably reliable correlation between core depth and age. Dating is now considered to be precise at least as far back as a hundred years.

The results from Colle Gnifetti leave no doubt that atmospheric pollution has reached even the highest peaks in the Alps. Sulphur concentrations, for example, have risen threefold during the last hundred years. Glaciers, it seems, have become witnesses to the sad history of atmospheric contamination.

Drilling through the ice sheets of Greenland and Antarctica

In 1956 and 1957 a team of scientists from America retrieved the first ice cores from the Greenland Ice Sheet, from depths of 305 and 411 metres. As drilling techniques then improved, progressively greater depths were reached subsequently, and in the 1960s cores gave us a remarkably detailed record of climatic evolution over more than 120,000 years, in other words, all the way through the last ice age and into the preceding warmer interglacial period. One of the most startling discoveries from these was that temperatures appear to change much more quickly than was previously thought. In human terms, the onset of an ice age may be considered a catastrophic event, rather than a gradual, imperceptible change in the climate.

The Greenland ice cores have also yielded evidence of some of the causes of climatic change during the past 10,000 years, that is since the last ice age. Acidity in the ice, mainly due to sulphuric acid from volcanic fall-out, has been measured by Danish scientists testing the electrical conductivity of melted ice samples to record evidence of major volcanic eruptions. Various Icelandic eruptions are well represented in the samples, but even the disastrous Krakatoa eruption of 1783 in far-distant Indonesia left its mark. One of the cores also recorded the eruption of the volcano Thera on the island of Santorini in the Aegean Sea, which is believed to have caused the disappearance of the Minoan civilization. By correlating the volcanic acidity record with a Northern Hemisphere temperature index (based on climatic records from England, tree ring widths from California, as well as the variation in precipitation on the Greenland Ice Sheet), the Danish scientists were able to demonstrate that major eruptions have a considerable effect on climate. This, of course, further complicates predictions about the climate's future behaviour.

What is probably the oldest ice on Earth has been recovered from the East Antarctic Ice Sheet. Here, at the Soviet research station of Vostok (which, incidentally, boasts another record – that of being the world's coldest place), an ice core 2000 metres long has recently been recovered. This takes us back, not only to the last interglacial period, but even into the preceding ice age. This core has taught us, amongst other things, that cold phases in the Earth's climatic history seem to coincide with very low levels of carbon dioxide concentrations in the atmosphere. However, we might be considering a chicken-and-egg situation. On the one hand, did the carbon dioxide become reduced first (and if so, why?), thereby reducing the atmosphere's natural greenhouse effect, and thus initiating an ice age? Alternatively, did lower temperatures (caused by other factors) lead to the reduction of the carbon dioxide? We do not know the answer, but clearly a close relationship between greenhouse gases and changes in climate has been established.

A 40-million-year record from Antarctic sediments

Although ice cores can provide a detailed and continuous record of climate changes over many tens of thousands of years, the full climatic

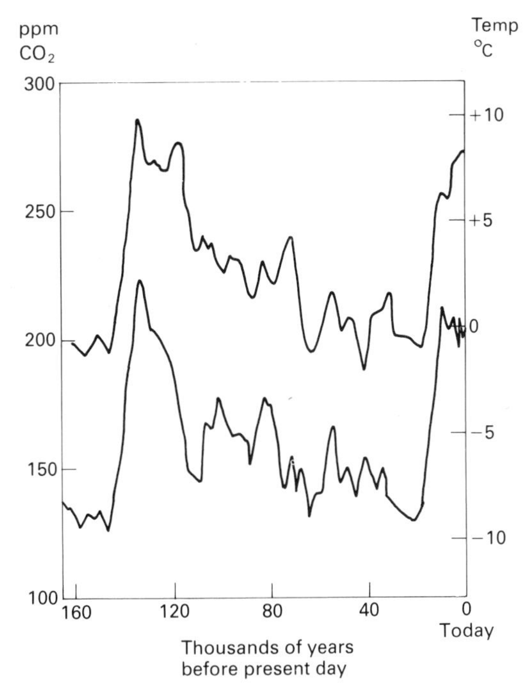

Summary of the 160,000-year-long climatic record from the Vostok ice core, East Antarctica. The upper curve and left-hand scale show the measured levels of carbon dioxide in the ice. The lower curve and right-hand scale indicate the temperature difference from today's levels, as determined from the isotopic ratios of the ice. Together they clearly indicate that high carbon dioxide values are associated with high temperatures.

On 30 August 1965, during the construction of a hydro-electric dam at Mattmark, a million cubic metres of ice broke off the tongue of Allalingletscher and buried a construction workers' camp. Eighty-eight people lost their lives. This catastrophe stimulated considerable glaciological research into the stability of steep ice masses and the dynamics of ice avalanches, in the hope that improved understanding of the processes involved might help to prevent similar occurrences in the future.

history since ice was established on earth during the current glacial era can only be determined from sediments in the seas surrounding the ice sheets. We now know from offshore sediments that the Antarctic Ice Sheet has been in existence for at least forty million years, whereas the ice sheets in the northern hemisphere only became established $2\frac{1}{2}$ million years ago.

Two major drilling operations have been carried out on the Antarctic continental shelf, in which one of the authors participated. In late 1986, the New Zealand Antarctic Research Programme drilled to a depth of 702 metres below the sea floor of McMurdo Sound in the western Ross Sea, using the winter sea ice as a drilling platform. It recovered a remarkable record of sedimentation from icebergs, floating ice and grounded ice spanning 36 million years, including a period that once was thought to have been ice-free. Even more surprising was the evidence for extensive ice at sea level when scrubby beech woodland was growing onshore in ice-free enclaves, demonstrating that a major ice sheet could then survive much higher temperatures than those of today.

The second major drilling operation was undertaken by an international team of scientists during the ship-borne Ocean Drilling Program to Prydz Bay, East Antarctica in early 1988. This project confirmed the existence of a major ice sheet, with icebergs calving at sea level, as far back as 40 million years ago. At several times in the ice sheet's early history, ice advanced further over the continental shelf than the present day, yet there is a general consensus from other evidence that the temperatures globally were higher at that time. One plausible explanation for this is that higher temperatures lead to more moisture being picked up from the sea and deposited as snow on the continent. Thus precipitation, rather than temperature, is the key factor controlling the growth and decay of the ice sheet. The implication of this is that it is dangerous to assume, as many people do, that global warming will lead to the melting of ice sheets. The continental shelf sedimentary record, when properly dated and fully understood, is likely to provide useful constraints on theoretical models of ice sheet growth and decay.

One of the major problems of the continental shelf sedimentary record is that it is difficult to date because there are so few fossils. Furthermore, the record is incomplete, as successive glacial advances have

removed some of the earlier sediments. A more continuous record, however, is available in the deep ocean basins, where numerous investigations have been made under the Ocean Drilling Program and its predecessor, the Deep Sea Drilling Project. Small calcareous fossils called foraminifera, which often make up the bulk of the sediment, have been removed and subject to detailed isotopic analysis. In particular, oxygen isotopes, as in ice cores, provided an indication of the former temperatures, and indirectly global ice volume changes. These sediments can also be dated accurately, so quite detailed temperature curves can be obtained. The main problem is that the precise role of ice volume changes is difficult to assess and scientists do not always agree on how to interpret the data.

The combination of all these methods is providing us with an increasingly detailed picture of climatic change on different time scales. The human historical time-scale is represented by the high Alpine cores on Monte Rosa, the time-span of a complete glacial-interglacial cycle by the ice cores from the polar ice sheets, and the time-span of a complete glacial era by the deep sea and continental shelf sediment cores. All approaches are necessary as we try to unravel the secrets of the earth's climate system.

Catastrophes

Over the centuries glaciers have been responsible for their fair share of disasters. We have already described the perilous combination of glaciers and volcanoes (chapter 7), but there are other ways in which they can cause havoc, notably by ice avalanches and bursts from glacially dammed lakes. Over the centuries, the Swiss Alps have had their share of such disasters, although in terms of the number of fatalities they have been small compared with those in more geologically unstable areas, such as the Andes.

Ice avalanches

'On August 31, 1597 a mass of ice detached itself from the Balmengletscher, fell onto the village of Eggen and buried it together with 81 people, all cattle and all other possessions. Nothing could be saved.

Fortunately, many inhabitants at that time were still working on the alpine pastures. Otherwise the peril would have been even greater. The ice mountain took seven years to melt away . . .'. This is the earliest historical record of a catastrophe caused by an ice avalanche. The village, then situated near the Simplon Pass in the canton of Valais in Switzerland, no longer exists, and neither does the glacier, having disappeared this century as a result of climatic warming.

However, the threat of ice avalanches from glaciers persists, both in the Alps and in other densely populated, glacierized mountain ranges. The Valais, in particular, has had a sad history of other ice avalanche catastrophes. One of the worst occurred on 30 August 1965 during the construction of a dam for a hydro-electric power plant at Mattmark in the Saas valley, when a mass of ice, with a volume of about a million cubic metres, broke away from the tongue of the Allalingletscher. Within seconds the avalanche had swept over part of the construction site burying the workers' camp and killing eighty-eight people.

The Mattmark ice avalanche triggered research into the mechanism of ice avalanche formation, and the distance an ice avalanche is likely to travel, that is its 'run-out distance'. A major problem in predicting ice avalanches is that, despite their spectacular effects, they are relatively rare, much rarer than snow avalanches. Nevertheless, dozens of large ice avalanches in the Alps and North America have been detected on thousands of aerial photographs taken especially for this purpose, and the data resulting from the mapping of these avalanches has then been used to predict, at least approximately, run-out distances.

However, predictions as to the distance an ice avalanche will travel remain difficult and notoriously inaccurate. This is partly because the roughness of the terrain has a complicated and variable braking effect on the falling ice mass. For example, the avalanche path may be very rough in summer and slow the avalanche down more effectively than in winter, when the path is covered with snow. In addition, the total volume of ice is usually very poorly known. Often, the ice does not break off the glacier in one piece, and smaller chunks may fall off weeks before the major event. One particular recurring case, the Balmhorngletscher in Switzerland, has become a tourist attraction, because of its repeated generation of ice avalanches down a steep rock face.

Right A curious block of ice has fallen off the southern hanging glacier of the Mönch in the Bernese Oberland, Switzerland. Despite its moderate mass of about 55 tonnes, the relatively compact firn surface has allowed it to travel an exceptionally long distance – 470 metres – down Grosser Aletschgletscher. Falling ice is one of the dangers that climbers face in glacierized mountain areas.

Below A much larger ice mass has here fallen off the Grosser Aletschgletscher. More than 330,000 cubic metres of ice stopped only about 70 metres short of the summer ski-lift visible in the lower right of the picture. Fortunately, the avalanche came down at night, traversing a frequently used route from Jungfraujoch to an Alpine Club hut. A new track skirts the avalanche on the right.

Above On 31 May 1970 one of the largest ice avalanches ever recorded combined with a rock fall, and descended from Nevado Huascaran Norte, the left-hand summit above, killing about 15,000 people in the town of Yungay. The avalanche track can still clearly be seen in this 1980 photograph.

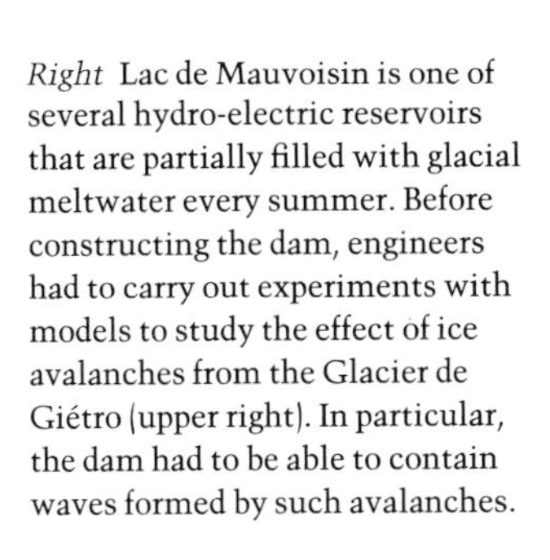

Right Lac de Mauvoisin is one of several hydro-electric reservoirs that are partially filled with glacial meltwater every summer. Before constructing the dam, engineers had to carry out experiments with models to study the effect of ice avalanches from the Glacier de Giétro (upper right). In particular, the dam had to be able to contain waves formed by such avalanches.

A series of photographs showing the collapse of part of the advancing Festigletscher on 18 August 1981. About 2000 cubic metres of ice were involved in this avalanche. Festigletscher lies above the village of Randa in the Valais, Switzerland, and is well seen from the nearby Alpine Club Domhütte.

In some instances a glacier may provide no prior warning of an impending catastrophic ice avalanche. In 1895 almost an entire glacier slid off the steep north-western face of Altels. More than four million cubic metres of ice fell down, and six shepherds and many cattle on an alpine meadow were killed. Calculations have shown that the ice must have reached a speed of 400 kilometres an hour as it swept across the route to the Gemmi Pass, some of it ending up nearly 400 metres above the valley floor on the opposite side. A similar event had occurred 113 years earlier.

Glaciologists have tried to find ways of predicting the ice volume to be released in avalanches, as well as the time of the event. It is now known that the ice in the unstable part of a glacier accelerates drastically prior to break-off, usually (but unfortunately not always) creating fresh crevasses. But in practice it is very difficult to monitor these developments, since they occur most frequently at high altitudes. Further research is essential, since mountain regions like the Alps are increasingly being used for recreation, with the establishment of ski resorts and transport routes, and for the generation of hydro-electric power. For example, when a hydro-electric reservoir was proposed at Mauvoisin in the Val de Bagne in Valais, it was known that ice avalanches from Glacier de Giètro could reach the site. Estimates were made of the waves that might be created by an avalanche falling into the water, and the dam was constructed high enough above the normal water level to contain them.

Glacier-related disasters in Switzerland (excluding those involving mountaineers)

Year	Location	Event	Fatalities
1595	Glacier de Giétro	Lake outburst	160
1597	Balmengletscher	Ice avalanche	81
1636	Weisshorn	Ice and snow avalanche	37
1782	Altels	Ice avalanche	4
1818	Glacier de Giétro	Lake outburst	40
1819	Weisshorn	Ice and snow avalanche	2
1895	Altels	Ice avalanche	6
1965	Allalingletscher	Ice avalanche	88

In December 1941, 6000 people, one third of the population of the town of Huaraz in the Peruvian Andes, drowned in a flood caused by an outburst of the glacial lake Laguna Palacoche. A very similar lake above the same city, albeit one which has not yet burst, is seen in this 1980 aerial photograph. The lake came into existence when the valley glacier in the middle of the picture retreated as a result of the world-wide twentieth-century warming. The lake level has been lowered artificially in order to prevent another catastrophe.

Ice avalanches are a threat to mountain settlements in glacierized mountain regions throughout the world. The valley of the Rio Santa in the province of Ancash in the Peruvian Andes, experienced one of the worst known glacier disasters. This beautiful valley lies at the foot of the Cordillera Blanca, an extensively glacierized range with dozens of peaks towering above 6000 metres. The highest of these, Nevado Huascaran (6768 metres), is also the most unstable mountain in Peru. Twice this century colossal ice and rock masses have become detached from the precipitous north summit: in 1962 about 4000 people were killed in an ice avalanche, and only eight years later, on 31 May 1970, an even worse disaster struck, when an earthquake of magnitude 7.8 on the Richter scale triggered a second, much larger avalanche. Rock from the west face of the mountain broke off, carrying with it part of the glacier that it supported, and the whole mass shot downvalley. Approximately 50 million cubic metres of ice, rock, morainic debris and water covered the 16 kilometres to the valley bottom in about three minutes. This avalanche spread much further laterally than the earlier one, burying the town of Yungay. About 18,000 people were killed there alone, and in all the earthquake claimed a total of 70,000 lives.

Outbursts from glacial lakes

Glacier and snow melt sometimes combine with rainwater to form ice-marginal lakes, such as where an ice-free side valley enters the main glacier-filled valley, or at the meeting place of two glaciers. Because internal plumbing is less efficient in cold glaciers, large ice-dammed lakes are more common in polar regions. But wherever they form, the pattern is the same: they tend to fill up gradually during the summer until a sufficient head is generated to make the glacier lift up from its bed, allowing the lakes to empty dramatically and producing large floods downvalley. Alternatively, the lake overflows, thence cutting rapidly down through the ice, again producing a flood. The complete cycle of filling and emptying may occur annually, but outbursts may also happen unpredictably.

Although ice-dammed lakes in temperate latitudes tend to be smaller than those in polar regions, their proximity to centres of population can cause considerable damage when they burst.

The biggest glacier-dammed lake in Peru is Laguna Parron. Its waters are held back by a steep, debris-covered glacier which descends from Nevado Hunadoy. In order to protect several towns below the lake from potential outbursts, Peruvian and French engineers lowered the lake level by several metres by drilling a tunnel through the mountain flank on the lower left of the photograph.

We have already described subglacial lakes in reference to volcanic activity but they can occur under any temperate glacier or under cold glaciers that have been warmed up at their base by geothermal heat to the melting point. Indeed, the biggest subglacial lakes are probably those beneath the Antarctic Ice Sheet. They have never been seen, but were discovered during the course of radio-echo sounding surveys designed to determine the thickness of the ice sheet. The lakes being hidden from sight, outbursts from them are totally unpredictable.

The worst recorded glacier lake outburst disaster took place in the

same mountain valley in Peru that experienced the terrible ice avalanches referred to above. On 15 March 1941 the busy market town of Huaraz was partially destroyed by a flood and 6000 people drowned. This outburst was from a sort of glacial lake that is characteristic of the Cordillera Blanca. Numerous terminal moraines were left behind in this range as glaciers retreated during the first two thirds of this century. Many of them formed natural basins which fill with rain and meltwater. Because they contain loose debris the moraines are easily eroded, particularly during thunderstorms or at times of strong meltwater discharge from the glaciers, and as an outlet channel deepens, increasing volumes of water rush through it, thereby eroding the moraine at an ever-increasing rate, until the lake literally bursts out of its confines.

Peruvians have made strenuous and demanding efforts to remove some of the danger spots. They have transported equipment up to altitudes reaching above 4000 metres in order to construct artificial channels or, in the case of Laguna Paron, the largest of these lakes, to drill a tunnel into the lake bed from below. Such work is expensive, especially for a country stricken by poverty and political instability.

Outbursts from glacial lakes and of water from within glaciers are more frequent and widespread events than ice avalanches. Repeated and catastrophic floods have been recorded since 1788 in the Mendoza valley in the Argentinian Andes, and since 1600 in the Oetztal in the Austrian Alps. In both cases a glacier surged out of a tributary valley and entered the main valley, damming the river and creating a lake. Later, the water flowed over or broke through the ice dam. Glacier surges in the Mendoza valley recur about every half century, each time causing a flood; the latest was in 1985. On the other hand, the glacier Vernagtferner, responsible for the Austrian floods, has receded to such an extent that no surges have been observed in it this century.

Overleaf Irrigation channels in the Rhône valley, the driest place in the Swiss Alps, carry glacial meltwater to fields and pastures. The farmland is not only irrigated thereby, but also provided with mineral-rich sediment, thus improving crop yields.

The Benefits of Glaciers

Mark Twain tells how he once travelled to Zermatt, took the mountain railway up to Gornergrat, a well-known viewpoint from which tourists can observe the Matterhorn, and then sat down on nearby Gornergletscher. He expected to be transported by the glacier's motion down the

valley, back to Zermatt, thus saving the return ticket on the railway. Obviously, the story is a parody and Mark Twain surely knew that glaciers do not move very far in just a few hours. But could he have dreamt that in the future glaciers would indirectly provide power for the Swiss railways?

Earlier in this chapter we saw that glaciers can create a serious hazard in mountain regions by producing ice avalanches and floods. On the other hand, glaciers are also an invaluable asset to man and usually the benefits they provide far outweigh the disadvantages.

Irrigation and energy supply

A number of the world's desert regions, such as north-western China, the Thar desert of north-western India and Pakistan, the coastal desert of Peru or the wine-growing area of Mendoza in Argentina, all receive waters from adjacent mountain ranges. For example, Himalayan meltwater and snowmelt are fed into the Rajasthan canal system, its hundreds of kilometres of irrigation channels bringing life to the Thar Desert. Some of this water is glacial runoff – perhaps not a large contribution, but one that is greatest when it is needed most: in hot, dry years.

Perhaps surprisingly, glacial meltwater is important for irrigation in the Alps. Central valleys, such as the Rhône in the canton of Valais, Switzerland, are protected by mountain ranges to such an extent that water shortage limits the productivity of the land. However, plenty of rain and snow fall on the mountains, and over hundreds of years an intricate network of irrigation channels has been built and maintained to bring this to agricultural areas. To get the water there some of the channels have had to be installed in unusual ways in out-of-the-way places; for example, some made of wood have been fixed to vertical rock faces. Glacial meltwater in these channels carries fine sediment which is useful not only in making holes in the channels water-tight, but also in providing a rich supply of minerals to the pastures and fields that are being irrigated.

Agriculture has never been big business in alpine regions, but hydroelectric power generation is. Switzerland's hydro-power stations were built mainly in the 1950s and 1960s, and use most of the available water.

Dams hold back the summer's meltwater when electricity consumption is relatively low, but in winter, when demand is greatest, half of Switzerland's energy production is generated by water released from the reservoirs. The Massa hydro-electric power station near Brig, owned by the Swiss Railways, runs almost entirely on meltwater from the Grosser Aletschgletscher, the largest glacier in the Alps. Again water is most abundant in warm, dry years, and at such times glaciers provide the bulk of the energy needed to operate the Swiss railway system.

Glaciers do not always co-operate with the engineers who tap their water. In several cases glacier tongues have advanced over the tunnel intakes through which meltwater is transferred into the reservoirs. The intakes then become inaccessible from the surface and glacial sediment blocks them. Following the loss of intakes at the Allalingletscher near Saas Fee, engineers decided to install intakes in front of the rapidly advancing Findelengletscher near Zermatt in such a way that ice could flow over the site without closing it off. However, the glacier's advance has since slowed down, and so far the usefulness of the installations has yet to be proved.

Since glaciers provide a steady water supply during long periods of summer drought, they are already widely used as a water resource for hydro-electric power generation in several temperate mountain countries other than Switzerland. Management of glacial meltwaters is already highly developed in Norway, and there is considerable scope in the more impoverished lands of the Himalayas and Andes.

Tourism

Some of the major tourist resorts such as Zermatt and Saas Fee depend considerably on their neighbouring glaciers. Apart from the attractive scenery, the Theodulgletscher near Zermatt and the Feegletscher above Saas Fee provide extensive areas of relatively crevasse-free firn which have been developed for summer skiing. The firn basins are high enough to provide dozens of kilometres of piste skiing even in July and August, thus making the installation of lift facilities economically viable, although special technology has had to be developed in order to anchor the skilift pylons effectively on the moving ice. In years when winter snowfalls are limited, such as those of the late 1980s, and the traditional

La Paz, the capital of Bolivia, is the highest city in the world with a population of more than one million (altitudinal range 3200 to 4100 m). It depends considerably on glacial meltwater for its water, since it regularly experiences long dry spells. Mururata, one of the mountains in the Cordillera Real, is visible in the background, bearing a large plateau glacier.

The Oberaargletscher provides water for a dammed lake, the Obersee, that is part of a complex hydro-electric scheme in Haslital in the Bernese Oberland, Switzerland.

skiing areas are starved of snow, the summer ski areas are put to good use in winter too.

Other, but smaller, skiing areas have been established on glaciers in several other resorts in the Alps. From the glaciologist's point of view, a particularly interesting case is the Titlis mountain near Engelberg in central Switzerland. A spectacular piste on Titlisgletscher provides good skiing until quite late in spring or early summer, but in its natural state it would not be suitable. However, a steep, icefall-like section of the glacier is extensively prepared every autumn, before the skiing season starts. Caterpillar tractors are used to fill crevasses with ice debris and a terrace-like route is blasted across a particularly steep ice slope. This is difficult because glacier ice is plastic and does not readily fracture, calling for special drilling techniques to place explosive charges at strategic points. Ice towers (séracs) are brought down slice-by-slice by repeated drilling and blasting.

Glaciers are tourist attractions all over the world. Large vehicles with snow tracks transport visitors around Athabasca Glacier in the Rocky Mountains of Canada. Frequent scenic flights with glacier landings are a major attraction in the Mount Cook region of New Zealand. The Bolivian city of La Paz boasts the world's highest ski area, on a small glacier more than 5000 metres above sea level in the Cordillera Real. An imaginative hotel owner in Grindelwald in the Bernese Oberland of Switzerland has installed a 'glacier clock', which entertains visitors by demonstrating the minute-by-minute motion of the glacier through a wire-and-pulley mechanism connected to the ice. And there must be dozens of artificial tunnels carved into glaciers around the world into which tourists can wander and experience the unique blue lighting characteristic of glacier ice.

What might a country with cold winters do if it doesn't have a glacier? It might adopt the scheme being mooted in Japan and manufacture one! Snow guns, already widely used in ski resorts, would provide the necessary additional accumulation, and it is anticipated that visiting tourists will make the venture financially worthwhile.

The commercial value of glacier ice

Glacier ice itself has proved to be a profitable export commodity for some countries. Ice exports were a feature of the Norwegian economy before refrigerators were invented, when 'ice houses' were common in the stately homes of Britain and France. Nowadays, the Japanese are able to cool their drinks with expensive ice hacked from an Alaskan glacier, while in Iceland melted glacier ice is sold as an unusually pure mineral water. In the Cordillera Blanca of Peru a few people still make a living out of collecting glacier ice, grinding it up, mixing it with flavouring, and then selling it at the markets as a local version of ice-cream. Donkeys and mules are used to transport the neatly cut blocks from nearly 5000 metres above sea level down to the valley; to slow ablation under the influence of the tropical sun and the animal's body heat, the ice is packed in thick layers of grass.

Longer-term possibilities exist for the large-scale use of glacier ice as a water resource, once the technological difficulties have been resolved and the economic conditions become favourable. The most promising possibility is the towing of Antarctic icebergs to areas with hot and arid climates, such as Australia, northern Africa or the Arabian peninsula, in order to provide clean drinking water, provided that berths can be found for objects having a draught in excess of the depth of the continental shelves. At times, Antarctic icebergs drift a considerable distance north under the influence of wind and currents, sometimes even as far as the Cape of Good Hope and Cape Horn, so the final towing distances may be less than at first expected.

What of the future?

The influence of glaciers on our lives cannot be underestimated, even if it is indirect. We do not know for certain what the future holds for Earth's ice cover, but there is no doubt that mankind is so radically polluting the globe that major changes of climate are to be expected.

Many scientists believe that the build-up of carbon dioxide will increase the so-called greenhouse effect of Earth's atmosphere so that it will become warmer. In polar regions, and especially the Antarctic, it is conceivable that large volumes of ice could melt.

Some scientists have suggested that the potentially unstable West Antarctic Ice Sheet could disintegrate catastrophically, leading to a

Skiers on a glacier at over 3000 metres in the popular ski region of Hintertux at the head of Zillertal, Austria, are served by a lift system that provides reliable snow in summer as well as winter.

global sea level rise of several metres within a time-span sufficiently short to make it difficult for human beings to adapt. However, the assumption is based on the ice shelves bordering the West Antarctic Ice Sheet becoming unstable, and as yet there is no on-site evidence to suggest that this will happen. It has also been suggested that the East Antarctic Ice Sheet has surge-type basins, and hence will transmit large amounts of ice into the southern seas, though the evidence for this seems to us dubious. On another tack, it is commonly assumed that global warming will result in melting of the Antarctic Ice Sheet and a rapid rise in sea level. However, it seems equally plausible that warming would lead to increased precipitation as snow and *growth* of the ice sheet. Basically, we have far too little hard data to make reliable predictions, hence there is a need for much more research.

It has been estimated that, *if* the entire Antarctic Ice Sheet were to melt, a volume of about 30 million cubic metres of water would be added to the oceans. This is equivalent to a sea level rise of about 80 metres, and would result in many of the world's major cities, such as London, New York, Buenos Aires, Calcutta, Shanghai and Tokyo being

largely submerged. This is most unlikely to happen, despite the spread of scare stories in recent years. However, it does show that even small percentage volume changes of the Antarctic Ice Sheet could have considerable effects on the sea level. It is quite likely, according to the available data, that a sea level rise of a metre over the next 100 years will take place. But in reality we do not know exactly what will happen. Some parts of the Greenland and Antarctic ice sheets are indeed shrinking, but other parts are growing, and the net change will only be determinable if more mass balance data become available, as is possible through new satellite facilities. Certainly, until we have a better understanding of what will actually happen to the ice sheets, it does seem sensible to insure ourselves against potential disasters by limiting the output of the various greenhouse gases through international agreement.

In at least one respect we know that pollution has caused severe damage to the atmosphere. Earth's ozone layer which provides a shield to ultraviolet radiation has been attacked by the widespread use of chlorofluorocarbons (CFCs). Although agreement has been reached to reduce the production of these chemicals, the growth of the Antarctic ozone hole and thinning elsewhere may be irreversible. Ozone depletion, and the associated increase in ultraviolet radiation at Earth's surface, may lead to marked climatic changes and the melting of ice sheets; other environmental damage may be even more severe, and there is already documentary evidence of increased skin cancer occurring in human beings.

At present we seem to be heading towards irreversible damage to Earth's ecosystem. International collaboration and vastly increased investment on research and education in environmental matters are necessary to reverse this trend, for we are running extreme environmental risks with little understanding of what the likely outcome will be. We do not even have a clear idea of what effect pollution has had on the climate in the past, although studies in the Alps and elsewhere are beginning to help.

Finally, glaciers are a recreational resource, favoured especially by skiers, mountaineers and mountain lovers. However, if they are not treated with respect by the individual, and protected from unsightly development by corporate bodies, we end up with unsightly concrete

Peruvian Indians pack ice blocks before transport from a 4800-metre-high glacier tongue to the town of Huaraz, where the ice will be ground, have flavours added to it, and be sold as ice cream. Straw mats help to reduce melting in the intense tropical sun.

and ironmongery despoiling the unique scenery, as has already happened in parts of the Alps, and may subject a public unaware of the hazards of glaciers to risk.

Whatever the short term effects of mankind's disruption of the climate, it is possible that glaciers will have the last word. We are living in an interglacial period within a long-running glacial era. In a few thousand years glaciers may once again extend equatorwards across the greater part of north-western Europe, North America and elsewhere, bulldozing cities, reducing growing seasons everywhere, destroying farmland, and making our current civilization impossible.

Glossary

Ablation The process of wastage of snow or ice, especially by melting.
Ablation area/zone That part of a glacier's surface, usually at lower elevations, over which ablation exceeds accumulation.
Accumulation area That part of a glacier's surface, usually at higher elevations, on which there is net accumulation of snow, which subsequently turns into firn and then glacier ice.
Arête (from French) A sharp, narrow, often pinnacled ridge, formed as a result of glacial erosion from both sides.
Basal sliding The sliding of a glacier over bedrock, a process usually facilitated by the lubricating effect of meltwater.
Basket-of-eggs topography Extensive low-lying areas covered by small elongated hills called drumlins (*q.v.*).
Bergschrund (from German) An irregular crevasse, usually running across an ice slope in the accumulation area, where active glacier ice pulls away from ice adhering to the steep mountainside.
Bergy bit A piece of floating glacier ice up to several metres across, commonly deriving from the disintegration of an iceberg.
Boulder clay An English term for till (*q.v.*), no longer favoured by glacial geologists.
Braided stream A relatively shallow stream that has many branches that commonly recombine and migrate across a valley floor. Braided streams typically form downstream of a glacier.
Breached watershed A short, glacially eroded valley, linking two major valleys across a mountain divide.
Calving The process of detachment of blocks of ice from a glacier into water.
Chattermarks A group of crescent-shaped friction cracks on bedrock, formed by the juddering effect of moving ice.
Cirque (from French) An armchair-shaped hollow with steep sides and back wall, formed as a result of glacial erosion high on a mountainside, and often containing a rock basin with a tarn (*q.v.*) (*cf. Corrie, Cwm*).
Cirque glacier A glacier occupying a cirque.
Col (from French) A high-level pass formed by glacial breaching of an arête or mountain mass.
Cold glacier A glacier in which the bulk of the ice is below the pressure melting point (although surface ice may warm up enough to melt in summer, while ice at the bed may also be warmed by geothermal heating).
Cold ice Ice which is below the pressure melting point, and therefore dry.
Compressive flow The character of ice flow where a glacier is slowing down and the ice is being compressed and thickened in a longitudinal direction.
Corrie (from Gaelic *coire*) A British term for cirque (*q.v.*).
Crag-and-tail A glacially eroded rocky hill with a tail of till formed downglacier of it.
Crescentic gouge A crescent-shaped scallop, usually several centimetres across, formed as a result of bedrock fracture under moving ice.
Crevasse A deep V-shaped cleft formed in the upper brittle part of a glacier as a result of the fracture of ice undergoing extension.
Crevasse traces Long veins of clear ice a few centimetres wide, formed as a result of fracture and recrystallization of ice under tension without separation of the two walls; these structures commonly form parallel to open crevasses and extend into them. Thicker veins of clear ice resulting from the freezing of standing water in open crevasses are also called crevasse traces.
Cryoconite hole A small cylindrical hole on the surface of a glacier, formed by small patches of debris that absorb more radiation than the surrounding ice, and melt downwards at a faster rate.
Cwm The Welsh term for cirque (*q.v.*), also sometimes used more generally outside Wales.
Dirt cone A thin veneer of debris draping a cone of ice up to several metres high, formed because the debris has retarded ablation under it.
Drumlin (from Gaelic) A streamlined hillock, commonly elongated parallel to the former ice flow direction, composed of glacial debris, and sometimes having a bedrock core; formed beneath an actively flowing glacier.
Englacial debris Debris dispersed throughout the interior of a glacier. It originates either in surface debris that is buried in the accumulation area or falls into crevasses, or in basal debris that is raised by thrusting processes.
Englacial stream A meltwater stream that has penetrated below the surface of a glacier and is making its way towards the bed.
Equilibrium line/zone The line or zone on a glacier's surface where a year's ablation balances a year's accumulation (*cf. Firn line*). It is determined at the end of the ablation season, and commonly occurs at the boundary between superimposed ice (*q.v.*) and glacier ice.
Erratic A boulder or large block of bedrock that is being

or has been transported away from its source by a glacier.

Esker (from Gaelic) A long, commonly sinuous ridge of sand and gravel, deposited by a stream in a subglacial tunnel.

Extending flow The character of ice flow where a glacier is accelerating and the ice is being stretched and thinned in a longitudinal direction.

Fault A displacement in a glacier formed by ice fracturing without its walls separating. It can be recognized by the discordance of layers in the ice on either side of the fracture.

Firn (from German) Dense, old snow in which the crystals are partly joined together, but in which the air pockets still communicate with each other.

Firn line The line on a glacier that separates bare ice from snow at the end of the ablation season.

Fjord (from Norwegian; **Fiord** in North America and New Zealand) A long, narrow arm of the sea, formed as a result of erosion by a valley glacier.

Fold Layers of ice that have been deformed into curved forms by flow at depth in a glacier.

Foliation Groups of closely spaced, often discontinuous, layers of coarse bubbly, coarse clear and fine-grained ice, formed as a result of shear or of compression at depth in a glacier.

Geothermal heat The heat output from the earth's surface. This affects glaciers especially in the polar regions, by warming the basal zone to the pressure melting point.

Glacial period/glaciation A period of time when large areas of the Earth (including present temperate latitudes) were covered with ice. Numerous glacial periods have occurred within the last few million years, and are separated by interglacial periods (*q.v.*).

Glaciated The character of land that was once covered by glacier ice in the past (*cf. Glacierized*).

Glacier A mass of ice, irrespective of size, derived largely from snow, and continuously moving from higher to lower ground, or spreading over the sea.

Glacier ice Any ice in, or originating from, a glacier, whether on land or floating on the sea as icebergs.

Glacier karst Debris-covered stagnant ice, sometimes found at the snout of a retreating glacier, with numerous lake-bearing caverns and tunnels.

Glacier sole The lower few metres of a (usually sliding) glacier that are rich in debris picked up from the bed.

Glacier table A boulder sitting on a pedestal of ice; the boulder protecting the ice from ablation during sunny weather.

Glacier tongue See *Ice tongue*.

Glacierized The character of land currently covered by glacier ice (*cf. Glaciated*).

Growler A piece of glacier ice almost awash, up to a few metres across, but generally smaller than a bergy bit (*q.v.*).

Hanging glacier A glacier that spills out from a high level cirque or clings to a steep mountainside.

Hanging valley A tributary valley whose mouth ends abruptly part way up the side of a trunk valley, as a result of the greater amount of glacial downcutting of the latter.

Highland icefield A near-continuous stretch of glacier ice, but with an irregular surface that mirrors the underlying bedrock and that is punctuated by nunataks (*q.v.*).

Horn A steep-sided, pyramid-shaped peak, formed as a result of the backward erosion of cirque glaciers on three or more sides.

Ice age A period of time when large ice sheets extend from the polar regions into temperate latitudes. The term is sometimes used synonymously with 'glacial period' (*q.v.*), or embraces several such periods to define a major phase in earth's climatic history.

Ice apron A steep mass of ice, commonly the source of ice avalanches, that adheres to steep rock near the summits of high peaks.

Ice cap A dome-shaped mass of glacier ice, usually situated in a highland area, and generally defined as covering up to 50,000 square kilometres.

Ice cliff (ice wall) A vertical face of ice, normally formed where a glacier terminates in the sea, or is undercut by streams. These terms are also used more specifically for the face that forms at the seaward margin of an ice sheet or ice cap, and that rests on bedrock at or below sea level.

Ice sheet A mass of ice and snow of considerable thickness and covering an area of more than 50,000 square kilometres.

Ice shelf A large slab of ice floating on the sea, but remaining attached to and partly fed by land-based ice.

Ice ship Pinnacles of ice, shaped like triangular sails, up to several metres high, formed as a result of differential ablation under strong solar radiation.

Ice stream Part of an ice sheet or ice cap in which the ice flows more rapidly, and not necessarily in the same direction as the surrounding ice. The margins are often defined by zones of strongly sheared, crevassed ice.

Ice tongue (or glacier tongue) An unconstrained, floating extension of an ice stream or valley glacier, projecting into the sea.

Iceberg A piece of ice of the order of tens of metres or more across that has been shed by a glacier into a lake or the sea.

Icefall A steep, heavily crevassed portion of a valley glacier.

Interglacial period A period of time, such as the present day, when ice still covers parts of the earth's surface, but has retreated to the polar regions.

Internal deformation That component of glacier flow that is the result of the deformation of glacier ice under the influence of accumulated snow and firn and of gravity.

Isotopes Varieties of elements, all with identical chemical properties, but not precisely the same physical ones.

Jökulhlaup (from Icelandic). A sudden and often catastrophic outburst of water from a glacier, such as when an ice-dammed lake bursts or an internal water pocket escapes.

Kame (from Gaelic) A steep-sided hill of sand and gravel deposited by glacial streams adjacent to a glacier margin.

Kame terrace A flat or gently sloping plain, deposited by a stream that flowed towards or along the margin of a glacier, but that was left above the hillside when the ice retreated.

Kettle (or kettlehole) A self-contained bowl-shaped depression within an area covered by glacial stream deposits, often containing a pond. A kettle forms from the burial of a mass of glacier ice by stream sediment and its subsequent melting.

Knock-and-lochan topography Rough, ice-abraded, low-level landscape, comprising small hills of exposed bedrock, and rock basins with small lakes and bogs.

Lahar Debris-flow consisting primarily of volcanic ash and lava boulders. Heavy rain and/or melting snow and ice during a volcanic eruption mixes with the loose deposits and forms fast moving tongues of slurry.

Little Ice Age The period of time that led to expansion of valley and cirque glaciers world-wide, with their maximum extents being attained in about 1700–1850 AD in many temperate regions and around 1900 in Arctic regions.

Mass balance (or mass budget) A year-by-year measure of the state of health of a glacier, reflecting the balance between accumulation and ablation. A glacier with a positive mass balance in a particular year gained more mass through accumulation than was lost through ablation; the reverse is true for negative mass balance.

Moraine Distinct ridge or mound of debris laid down directly by a glacier or pushed up by it. The material is mainly till, but fluvial, lake or marine sediments may also be involved. Longitudinal moraines include a **lateral moraine** which forms along the side of a glacier; a **medial moraine** occurring on the surface where two streams of ice merge; and a **fluted moraine** which forms a series of ridges beneath the ice, parallel to flow. Transverse moraines include a **terminal moraine** which forms at the furthest limit reached by the ice, a **recessional moraine** which represents a stationary phase during otherwise general retreat, and a set of **annual moraines** representing a series of minor winter readvances during a general retreat. A **push moraine** is a more complex form that develops especially in front of a cold glacier during a period of advance.

Moulin A water-worn pothole formed where a surface meltstream exploits a weakness in the ice. Many moulins are cylindrical, several metres across, and extend down to the glacier bed, although often in a series of steps.

Nunatak (from Inuit) An island of bedrock or mountain projecting above the surface of an ice sheet or highland icefield.

Ogives Arcuate bands or waves, with their apexes pointing downglacier, that develop in an ice-fall. Alternating light and dark bands are called **banded ogives** or **Forbes' bands**. Each pair of bands or one wave and trough represents a year's movement through the icefall.

Outwash plain A flat spread of debris deposited by meltwater streams emanating from a glacier (*cf. Sandar*).

Permafrost Ground which remains permanently frozen. It may be hundreds of metres thick with only the top few metres thawing out in summer.

Piedmont glacier A glacier which spreads out as a wide lobe as it leaves a narrow mountain valley to enter a wider valley or a plain.

Plastically moulded (or p-) forms Smooth rounded forms of various types cut into bedrock by the combined erosive power of ice and meltwater under high pressure.

Portal The open archway that develops when a meltwater stream emerges at the snout of a glacier.

Pressure melting point The temperature at which ice melts under a specific pressure. Pressure lowers the melting point below 0°C.

Randkluft (from German) The narrow gap that develops

between a rock face and steep firn and ice at the head of a glacier.

Regelation ice Ice which is formed from meltwater as a result of changes of pressure beneath a glacier.

Rejuvenated (or regenerated) glacier A glacier which develops from ice avalanche debris beneath a rock cliff.

Riegel (from German) A rock barrier that extends across a glaciated valley, usually comprising harder rock, and often having a smooth up-valley facing slope and a rough down-valley facing slope.

Roche moutonnée (from French) A rocky hillock with a gently inclined, smooth up-valley facing slope resulting from glacial abrasion, and a steep, rough down-valley facing slope resulting from glacial plucking.

Rock basin A lake- or sea-filled bedrock depression carved out by a glacier.

Rock flour Bedrock that has been pulverized at the bed of a glacier into clay- and silt-sized particles. It commonly is carried in suspension in glacial meltwater streams, which consequently take on a milky appearance.

Sandar (plur. Sandur) (from Icelandic) Extensive flat plain of sand and gravel with braided streams of glacial meltwater flowing across it. Sandur are usually not bounded by valley walls and commonly form in coastal areas.

Sea ice Ice that forms by the freezing of the sea (ice shelves and icebergs float on the sea).

Sedimentary stratification The annual layering that forms from the accumulation of snow, and that is preserved in firn and sometimes in glacier ice.

Sérac (from French) A tower of unstable ice that forms between crevasses, often in icefalls or other regions of accelerated glacier flow.

Sill A submarine barrier of rock or moraine that occurs at the mouth of, or between rock basins in, a fjord.

Slush avalanche (slush flow) A fast-flowing mass of water-saturated snow, most commonly occurring in early summer when melting is at its peak.

Snout The lower part of the ablation area of a valley glacier, often shaped like the snout of an animal.

Snow swamp An area of saturated snow lying on glacier ice.

Sole See *Glacier sole*.

Strain The amount by which an object becomes deformed under the influence of stress.

Striae Linear, fine scratches formed by the abrasive effect of debris-rich ice sliding over bedrock. Intersecting sets of striae are formed as stones are rotated or if the direction of the ice flow over bedrock changes.

Striated The scratched state of bedrock or stone surfaces after the ice has moved over them.

Subglacial debris Debris which has been released from ice at the base of a glacier. It usually shows signs of rounding due to abrasion at the contact between ice and bedrock.

Subglacial stream A stream which flows beneath a glacier, and which usually cuts into the ice above to form a tunnel.

Superimposed ice Ice which forms as a result of the freezing of water-saturated snow. It commonly forms at the surface of a glacier between the equilibrium line and the firn line, and gives the glacier additional mass.

Supraglacial debris Debris which is carried on the surface of a glacier. It is normally derived from rockfalls and tends to be angular.

Supraglacial stream A stream that flows over the surface of a glacier. Most supraglacial streams descend via moulins (*q.v.*) into the depths or base of a glacier.

Surge A short-lived phase of accelerated glacier flow during which the surface becomes broken up into a maze of crevasses. Surges are often periodic and are separated by longer periods of relative inactivity or even stagnation.

Tabular iceberg A flat-topped iceberg that has become detached from an ice shelf, ice tongue or floating tidewater glacier.

Tarn A small lake occupying a hollow eroded out by ice or dammed by a moraine; especially common in cirques.

Thermal regime That state of a glacier as determined by its temperature distribution.

Thrust A low-angle fault, usually formed where the ice is under compression. Thrusts commonly extend from the bed and are associated with debris and overturned folds.

Tidewater glacier A glacier that terminates in the sea.

Till A mixture of mud, sand and gravel-sized material deposited directly from glacier ice.

Tongue That part of a valley glacier that extends below the firn line.

Trim-line A sharp line on a hillside marking the boundary between well-vegetated terrain that has remained ice-free for a considerable time and poorly vegetated terrain that until relatively recently lay under glacier ice. In many areas the most prominent trim-lines date from the Little Ice Age (*q.v.*).

Tundra The zone of shrubs and other small plants that grow mainly on top of the permafrost in Arctic regions north of the tree line.

Unconformity A discontinuity in the annual layering in firn or ice, resulting from a period when ablation cut across successive layers.
Valley glacier A glacier bounded by the walls of a valley, and descending from high mountains, from an ice cap on a plateau, or from an ice sheet.
Warm glacier A glacier whose temperature is at the pressure melting point throughout, except for a cold wave of limited penetration that occurs in winter.
Warm ice Ice which is at melting point. The temperature may be slightly below 0°C at the base of a glacier where the ice is under high pressure.
Whaleback A smooth, scratched, glacially eroded bedrock knoll several metres high, and resembling a whale in profile.

Index

Numbers in italics refer to illustrations

Index